GENDERS AND SEXUALITIES IN EDUCATIONAL ETHNOGRAPHY

STUDIES IN EDUCATIONAL ETHNOGRAPHY

Series Editor: Geoffrey Walford

STUDIES IN EDUCATIONAL ETHNOGRAPHY   VOLUME 3

GENDERS AND SEXUALITIES IN EDUCATIONAL ETHNOGRAPHY

EDITED BY

GEOFFREY WALFORD

Department of Educational Studies, University of Oxford, UK

CAROLINE HUDSON

Department of Educational Studies, University of Oxford, UK

2000

JAI
An imprint of Elsevier Science

Amsterdam – New York – Oxford – Shannon – Singapore – Tokyo

ELSEVIER SCIENCE Inc.
655 Avenue of the Americas
New York, NY 10010, USA

First edition 2000

Library of Congress Cataloging in Publication Data
A catalog record from the Library of Congress has been applied for.

ISBN: 0-7623-0738-2

♾ The paper used in this publication meets the requirements of ANSI/NISO Z39.48-1992 (Permanence of Paper).
Printed in The Netherlands.

CONTENTS

vi

PREFACE

Ethnography has become one of the major methods of researching educational settings. Its key strength is its emphasis on understanding the perceptions and cultures of the people and organisations studied. Through prolonged involvement with those who are being studied, the ethnographic researcher is able gradually to enter their world and gain an understanding of their lives.

Each volume of *Studies in Educational Ethnography* focuses on a particular theme relating to the ethnographic investigation of education. The volumes are closely linked to an annual two day residential conference which explores various elements of ethnography and its application to education and schooling. The series of Ethnography and Education conferences began in the late 1970s, and was originally held at St. Hilda's College, Oxford University. The series later moved to Warwick University and back to the Department of Educational Studies, University of Oxford in 1996. Each year a broad theme for the conference is chosen and participants are invited to contribute papers. The conference meeting itself is a period of shared work: papers are pre-circulated to participants and critically yet supportively discussed. In their revisions for possible publication, participants are thus able to take account of the detailed critique offered by their colleagues.

The contributions presented in each volume of *Studies in Educational Ethnography* are of two types. Most are revised versions of papers presented at the annual Oxford Ethnography and Education conference, but each volume also includes some further specially commissioned pieces. They are selected on the basis of their high quality, their coherence as a group and their contribution to both ethnographic methodology and substantive knowledge.

The series recognizes that the nature of ethnography is contested, and this is taken to be a sign of its strength and vitality. While the idea that the term can be taken to be almost synonymous with qualitative research is rejected, chapters are included that draw upon a broad range of methodologies that are embedded within a long and detailed engagement with those people and organizations studied.

Further details of the Oxford Ethnography and Education conference or the *Studies in Educational Ethnography* series of volumes are available from the Series Editor.

Geoffrey Walford

Professor of Education Policy
Department of Educational Studies
University of Oxford
15 Norham Gardens
Oxford OX2 6PY
England

INTRODUCTION

Geoffrey Walford

This third volume of *Studies in Educational Ethnography* focuses on genders and sexualities in education. Studies of gender and education have been at the forefront of much ethnographic research for several decades, but the interest in sexualities and education is more recent. Indeed, to study sexualities and schooling is, almost in itself, a challenge to many common assumptions about children and what it is 'to be a child'. However, study of the nature and construction of gender and sexuality has now become a recognized and highly productive research area, and one that is highly suited to ethnographic methods. The new research presented in these chapters illustrates some of the range of recent work in the area and the power of ethnography to uncover the complex and contested processes of construction.

The first chapter by Tuula Gordon, Janet Holland and Elina Lahelma reports part of the findings from a comparative ethnographic study of four secondary schools – two in Helsinki and two in London. Following a description of their methods of data construction and analysis, they present material on the ways that girls interact with each other in the everyday life of the school. Their analysis conceptualizes the school in terms of the official, informal and physical layers and they have developed the metaphor of the dance to capture the relationship between these levels and to illuminate aspects of space, movement and embodiment. Throughout the account they move away from studies that emphasize conflict and resistance and illustrate the many cases of cooperation. They give examples of where girls reveal deep emotional bonds with each other, of mutual support against the activities of boys, and of how they negotiate gender relations together. The authors examine the contradictions between the supportive interactions that they have observed between

Genders and Sexualities in Educational Ethnography, Volume 3, pages 1–6.
Copyright © 2000 by Elsevier Science Inc.
All rights of reproduction in any form reserved.
ISBN: 0-7623-0738-2

girls, and the divisions and conflict that were prominently described by the girls in interviews and in other research. They argue for the importance of space in interpretation of data, as some spaces are safer than others for girls to work out their differences. They do not neglect the divisions and conflicts between girls, but recognize the importance of the researchers' personal experiences and political stances in the interpretation of data. They suggest that the support that girls give to each other is very important to them, and is particularly so if there are boisterous boys who demand disproportionate attention and energy from the teacher. Girls' cooperation makes the classroom a more pleasant place for them, and helps them to broaden the space they are using.

Mary Jane Kehily's chapter is concerned with the relationship between sexuality and schooling and draws upon data from an ethnographic school-based study. The chapter considers the ways in which sexualities are shaped and lived through pupil cultures which are often marginalized or overlooked by teachers and rarely find their way into the official curriculum. She then discusses the normative power of heterosexuality in schools and, particularly, the relationship between masculinities and heterosexuality. In this context, themes of embodiment, physicality and performance play a part in the ways in which informal groups of school students actively ascribe meanings to issues of sex and gender. Through interviews with young men in school, the chapter suggests that school processes produce sites for the enactment of heterosexual masculinities which suggest both the normative presence of heterosexuality and the fragility of sex – gender categories. As part of the recent encompassing of heterosexuality within academic debates, the chapter aims to contribute to an understanding of sexual majorities by focusing on the processes which are constitutive of dominant practice in school arenas.

The chapter by Andrew Parker also investigates heterosexual masculinities, but this time within the unusual setting of youth training within an English professional football club. Drawing upon his year-long ethnographic study, he presents a broadly descriptive account and analysis of how the everyday processes and practices within this particular form of youth training serve to shape the masculine identities of the trainees in accordance with both official and unofficial institutional norms, standards and expectations. Using the concepts of 'hegemonic masculinity' and 'working-class shop-floor culture', Parker describes the range of constructed institutional masculinities and categorizes the trainees according to their divergent lifestyle behaviours, their commitment to occupational success, and the specific masculine traits that they exhibit. He argues that the training scheme constitutes a *rite de passage* for these young men and is a pivotal stage in the masculine development of the individuals concerned.

In the following chapter Caroline Hudson draws on her ethnographic study of 32 young people to gain their accounts of their experience of family and schooling. She explores the relationship between these experiences and the diversity of family structures found in the group, and illustrates how gender might inform academic knowledge on the issue. She argues that much previous qualitative research on family structure has tended to stress differences in young people's experiences of family, according to family structure, and to underplay the potential influences of gender. Hudson argues that there is a need to recognize both differences and similarities according to family structure and gender, and shows that, whilst family structure was largely unimportant in these young people's accounts of the relationships with their mothers and fathers, gender was, in contrast, important. She shows how these young people tended to highlight close communication in their relationships with their mothers, and how this was particularly so amongst girls. Positive relationships with fathers, whether natural resident, natural non-resident, or step, tended to be characterized by shared activities, rather than by close communication. The case studies of three boys which conclude the chapter show how incorporating gender into an analysis of family structure can place in sharp relief potentially positive aspects of family reordering.

Christine Skelton's chapter presents some of the results from her year-long study of six to seven year-old boys in one infant class in a school sited in a disadvantaged area in the North of England. The chapter powerfully illustrates the ways in which masculine identity is a skilfully negotiated accomplishment drawing upon a range of discursive positions such as being a boy, white, child, school pupil, and a member of the so-called 'underclass'. Skelton describes some of the multiplicity of masculinites that the boys experience both inside and out of the school, and shows how these boys are forced to negotiate their own masculine identities. She examines the culturally exalted form of masculinity predominant in the area and the 'knowledge' that the boys thus brought with them to school regarding being a 'lad'; the discursive tensions between being a boy and a school pupil; and where the boys positioned themselves with regard to the 'apprentice lad' form of masculinity and their relationships with each other.

The following chapter by Geoffrey Walford focuses on the 'first days in the field' within an evangelical Christian school. This school is one of many small private schools in England that have been established because parents are dissatisfied with what they see as the increasing secularization of state-maintained schools. The school became a site for a compressed ethnography as part of a wider comparative study of government policy on schools for religious minorities, but gender and sexuality were two of the additional foci for observation. Indeed, Walford argues that it was difficult for him to avoid such issues, as many of the teachers saw issues connected to gender and sexuality as central to what differentiated their school from state-maintained schools. He shows that the secularization with which these parents and teachers were concerned has many facets, but the liberal view of sexuality that is believed to be propagated in many state-maintained schools was a key issue. For these Christians, sexuality is a gift from God that should not be abused, but reserved for heterosexual couples within marriage. Any homosexual activity was seen as particularly sinful. However, these first days in the field displayed some conflicting aspects to the school's understandings of issues of gender and sexuality. Some more liberal attitudes and behaviours were found within the dominant conservative stance, and there were some conflicting views. The chapter concludes with speculations on methodology and on substantive issues for the wider project.

Jayne Raisborough's chapter reports one particular aspect of her ethnographic study of the Sea Cadet Corps – a British uniformed youth organization with very close links to the Royal Navy. Raisborough's wider research shows how women in this organisation are expected and encouraged into labours and activities which accord with dominant heterosexual notions of femininity. Here, she focuses on the experiences of three lesbian officers and shows how they maintain an ambiguity which disguises their lesbianism but which challenges normative ideas of heterosexual femininity which underpin a gendered division of labour within the corps. She argues that discourses of political correctness can be used to compete with those of institutionalized homophobia to allow lesbian officers to move onto the *threshold* of the closet within which they remain. This position is shown to be an empowering place from which to dis-articulate homophobic and sexist voices.

Mary Jane Kehily and Anoop Nayak's chapter looks at the ways in which young women view the sex education they receive in school. This is considered in relation to their daily experiences of sexism and their negotiation of sexual relations. An ethnographic approach is used to illustrate the tensions between young women's social lives and the teaching and learning of sex education in the school curriculum. The chapter suggests that young women engage in sexual learning through informal sites such as friendship groups and popular cultural forms. Within this context magazines aimed at an adolescent female market and soap operas can be seen as cultural resources for teaching and learning about issues of sexuality. Finally, based on the views and experiences of the young women, the chapter suggests strategies for the change and development of sex education in schools.

The chapter by Valerie Hey is a more theoretical piece which takes her ethnography of girls' friendships as a starting point to re-think the concept of rapport and to investigate the politics of social class in feminist participant observation. In view of recent claims that the textual production of empathy is imperialistic, Hey argues from an alternative position, one that re/considers rapport as a contested feminist materialist praxis. In order to do this she revisits some of the questions that arose in conducting her ethnographic work in conditions of difference. Her objective is to seek the possibility of theorizing shared social experience and to rehabilitate possibilities for theorizing and/or recognizing sameness through a more historicized, psychically richer and socially saturated account of the notion of rapport. The chapter also reflects other cognate moves beyond the binaries of same/difference as these have emerged in feminist work on citizenship and post-colonial critical translations of literature.

The final two chapters are from one female and one male researcher who have been working on linked projects on the nature and effects of teacher stress. Denise Carlyle's research specifically focuses on emotion and stress-related illness and burnout among secondary school teachers. A study of such sensitive issues presents dilemmas for the researcher and those who volunteer to take part. Ten female and eleven male teachers were interviewed between two and nine times over a period of twelve months. A humanistic person-centred counselling framework informed by symbolic interactionism guided these in-depth life history interviews, and Carlyle was well aware that she was potentially disturbing a 'can of worms'. The lengthy conversations involved considerable discomfort, pain and soul-searching for both the participants and for the interviewer. These aspects are described and considered in the light of the researcher's own life history. She considers the various voices with which she conversed with the teachers, and focuses on the gendered voice, the academic voice and the teacher voice that she had presented. She argues that, while such interviews can be highly problematic, they can be growth-promoting and self-enhancing. She reports mainly beneficial effects.

The final chapter by Geoff Troman relates to the same overall teacher stress project, but his concern was with primary school teachers. His chapter gives a reflexive account of his experiences of interviewing and field relations, and he compares the experience of in-depth interviewing of male and female teachers who had recently experienced stress. The responses to the invitation to participate were such that his sample of 18 included 14 women. The discussion of the research experience is located in the methodological literature on gendered research where male accounts are still rare. In the chapter Troman gives an account of his own life history and his experiences of stress and relates this to the ways that he tried to develop trust and rapport with the teachers involved in the study. He then considers aspects of power relations within the interviews and the emotional aspects of conducting such interviews. He notes some key differences between the experience of interviewing women and men, showing, for example, that some of the interview sessions with women led to crying and acute embarrassment on behalf of both parties. This never happened with men. Troman speculates about the reasons for these differences, and ends with a recognition of the merits that a mixed-gendered team of researchers can bring.

FRIENDS OR FOES? INTERPRETING RELATIONS BETWEEN GIRLS IN SCHOOL

Tuula Gordon, Janet Holland and Elina Lahelma

INTRODUCTION

Norma goes up to Anja and flings her arms around her, with a big cuddle and giggles. In the games they are shouting out each other's names, and flinging arms around one another to celebrate. (ObsL).[1]

Girls are bent over their work, immersed in it, help each other. (ObsH)

I remember once that I had, we had, me and a couple of my friends had an argument with two other girls in our class and they spread this rumour around about us and they told somebody that we'd said something about that girl. Then this girl was like really popular and then like the whole class seemed to turn against us . . . we'd had this argument and they sort of thought that we meant something else by it, something bigger and so they thought, 'Oh, we have to get back at them' . . . and they started spreading this rumour. (FSL)

We are in three groups. Aura, Auli and Arja are usually together. Riikka and Inka, and then Marianne and myself. Then, in general, disputes between the groups occur. Mostly like, nobody really likes Riikka. Except Inka. And Aura is occasionally with her. Then, she's [Riikka] kind of like, I don't know, but kind of not so pleasant. (FSH)

The first two extracts from field notes are from lessons of 13/14 year-old students in a secondary school, the next two from interviews with young women from the groups that we followed. These contrasting comments illustrate the issues with which this chapter is concerned. The comments are drawn from the project 'Citizenship, difference and marginality in schools –

Genders and Sexualities in Educational Ethnography, Volume 3, pages 7–25.
Copyright © 2000 by Elsevier Science Inc.
All rights of reproduction in any form reserved.
ISBN: 0-7623-0738-2

with special reference to gender', a detailed, collaborative ethnographic study in which everyday processes in four schools, two in Helsinki and two in London, were compared. We were interested in axes of differentiation and our starting point was gender, but we saw gender as intersecting with other differences of social class, 'race', ethnicity, sexuality and impairment and all formed the basis for our observations, and the multiple methods of data collection we employed. We were interested in how processes of inclusion and exclusion in schools produce the citizen of the nation state and how identity and subjectivity are forged in these processes.

The project was located within the context of New Right policies in education and their relationship to equal opportunity politics. We compared Finland with its social democratic welfare state, and the British model which is based on conservatism and liberalism, but includes social democratic and radical elements. In education, at the time of the study, the two countries were moving in opposite directions, the U.K. taking more centralized control over the curriculum and reducing teacher autonomy by introducing a National Curriculum, and Finland loosening the tightly controlled National Curriculum and giving schools and teachers some autonomy. But the influence of New Right policies has been felt in each and the effect has been to move the debate on aspects of the welfare state, particularly education, to the right, and to embed notions of the market into the system. The setting of this agenda has been so effective, that despite changes of government in both countries, many of the policies and practices of the New Right have been incorporated by the new governments.

As a result of these changes in the educational agenda, in current education policies standards and choice are juxtaposed with equality; but historically and more generally schools and education have been expected both to confirm and to challenge social divisions, to promote emancipation and social mobility whilst at the same time engaging in regulation and the maintenance of stability. These continuing and contradictory expectations can be observed in school processes in tensions between agency and control. Our concern was to examine not only what could be seen as the repressive aspects of school required by their task of maintaining stability and social order, but to consider cooperation as well as conflict in school processes and interactions. As does Giddens (1985) we regard every limitation as providing an opportunity for enactment, every attempt at control as offering opportunities for agency. It was in this connection that the contrast between the cooperation which we noted amongst girls in the classroom observations and the conflict between them which frequently emerged in our interview data came to our attention.

OUR ETHNOGRAPHIC APPROACH

Our approach is influenced by Sara Delamont & Paul Atkinson (1995) who advise school ethnographers to become 'promiscuous bricoleurs, selecting whatever techniques, theories, or insights can be best deployed in any particular project' (p. vi). They urge ethnographers to take an eclectic and pragmatic approach to research and not see methods, disciplines and schools of thought as "sectarian doctrines with iron barriers between them." In developing our conceptual and theoretical framework we have drawn freely on a range of theoretical perspectives, including social constructionist, cultural, materialist, poststructural and feminist theory. The sociological roots of social constructionism are longstanding, but recent social constructionist approaches emerge from the desire to theorize individuals as agents with subjectivities in the production of society as well as to consider the constraints and possibilities with which society confronts the individual, and this is important in our approach. Burr (1995) suggests three central tenets of social constructionism: challenging taken for granted knowledge; historical and cultural specificity; recognition that knowledge is sustained by social processes and that knowledge and social action go together. These elements recur in other theoretical perspectives upon which we draw.

We retain an element of materialism, in terms of both the material base for social class and other differences upon which human agency is built. We look at processes and practices in everyday life at school by analyzing the wider social and historical relations in which these are expressed; Angela McRobbie (1996) calls this type of approach "new ethnography" in contrast to textual anti-essentialist, anti-experience approaches (McRobbie, 1997). Materiality is particularly important for us in relation to our interest in spatiality and embodiment. Space is social and mental, and constrained but not determined by the physical. The material body is given meaning through processes of social construction; ideas about the body are social, but they are not entirely separable from bodily constraints and possibilities.

These concerns form the basis for our analysis and interpretation, and have led us to make an analytical distinction between three layers of the school which formed the focus of our interest: the official, informal and physical school. The *official* school is laid out in documents of the school and the state; in our observations in the classroom we focus on lesson content, text books, teaching materials and methods, and classroom interaction. We examine the disciplinary apparatus of the school, rules and sanctions, and outline official hierarchies among and between teachers and students.

We chose to use the term *informal* to avoid a binary opposition between the official and the unofficial, to indicate that the informal school is different from, and not merely a reaction to, the official school, and that it has a life and a meaning of its own. Here we expand the analysis of classroom interaction to examine interaction between teachers and students beyond the instructional relationship, among students in other areas of the school, among teachers, and between teachers, students and other groups in the school (support staff of various types). School rules are compared with their enactment in practice, informal hierarchies among and between teachers and students are compared with formal hierarchies.

Our focus on the *physical* school, with the possibilities and limitations offered by school buildings and spaces for teaching and learning in the official school, and interactions and hierarchical differentiation in the informal school, was innovative at its inception. Here we draw on Giddens's (1985) discussion of time – space paths. We contrast rules about movement, talk, noise and the use of spaces in the school with the informally sanctioned and forbidden practices of students. The school as a physical space provides a context for the practices and processes that take place within it, but it is more than context, it shapes these practices and processes, and in this way can produce differentiation between students.

We have developed a metaphor to capture the relationship between the official, informal and physical layers of the school which also captures our interest in space, movement and embodiment – that of the dance.[2] Dance, Maureen Molloy (1995) suggests, is "structural and processual, involving set moves and individual improvisation, group patterns and individual performance" (p. 108) and, usually, both women and men. Here the *physical* is the dance hall or dance floor, the space in which the dance takes place and in which bodies move, as well as the dress and other embodiment of the dancers; the *official* is the rules of the dance, the formal steps, movements and dress codes prescribed for the specific dance (for example the minuet);[3] and the *informal* is all the exchanges that take place around and between these two elements (verbal, physical and other contacts).

DATA AND ANALYSIS

The authors, augmented by four researchers in Helsinki and two researchers in London,[4] gathered data in four secondary schools, one predominantly middle class and one predominantly working class in each of the cities. All the researchers were committed to collective, feminist research, and we wanted to ensure that we undertook collaborative work in which the difficulties of such an undertaking were turned into strengths. Our methods were observation, participant observation, formal and informal interviews, metaphors and associations. A comprehensive observation schedule was developed for each layer of the school, the official, the informal and the physical, to direct the gaze of the researcher onto as many relevant aspects of those layers as possible. Each researcher followed one class in each school, but members of the Helsinki team visited each other's classes regularly. With the schedules in their heads, researchers took copious notes on what they saw, sometimes feeling that our mode of observation reflected that of the ideal student, more specifically the ideal female student: quiet, industrious, observant, diligent, writing. Feelings, too, were important, as part of what we saw, and experienced. An observer's notes recall the range of feelings that we recorded during our stint in the classroom:

> ... anger, boredom, disappointment, sometimes even fear, loneliness, ambivalence, uncertainty, joy, excitement, tenderness, disgust, hurt, intolerance, sense of injustice. Not to mention headaches, sore wrists, stiff necks, aching backs, oxygen-deprived brains ...

The Finnish project team met regularly, discussing the observation experience and records in great detail, enabling them to hone and develop the way in which the observations were conducted. The collective work enabled the development of a shared language for observations and interpretation. The researchers in London followed this same practice amongst themselves, and the author here did so in interaction with the Finnish authors, by email, telephone, and face-to-face contact. Both within and between the London and Helsinki teams, a practice of analysis through discussion was generated; particular interactions, incidents or trends were chosen for joint focus, with the aim of grasping both general and particular links. These processes have been full of analytical insights, and it has proved to be very challenging, and at times impossible, to try to develop ways of writing such analyses. The Helsinki team followed together more than 900 lessons and numerous other occasions in the schools, for example staff meetings, parents' evenings, extra days and discos. The London researchers followed about 320 lessons, plus assemblies and other events in the school, and elsewhere occasionally when the class was on the move.

In the analysis of the data, all of the lesson notes were read, but in each case a time period was selected from the large number of lessons observed for detailed analysis. The material had three analytic readings. The first was thematic – where we looked for the major themes emerging from the data; the second was conceptual/interpretive – where we made interpretations on the basis of concepts we had developed; and the third was extractive – where we drew out illustrations of themes and concepts.

We interviewed teachers, support staff and all the students in the classes we followed, where possible.[5] Interviews were given an overall reading, highlighting areas of interest within the framework of the study. This was followed by an interpretive, conceptual reading, from which we generated files referring to marginality, difference, individuality and nationality. A further content analysis was undertaken by the Finnish researchers to generate a set of descriptive codes based on the interview content, and these were used to code the interviews on NUD*IST. While these codes were essentially descriptive, some of them intersected with, and were incorporated into, the more thematic and conceptual codes developed in other readings.

Teachers and students completed a short 'metaphor questionnaire' which involved questions about what they did in the lunch break, spaces around the school that they liked or disliked, and required them to produce a metaphor for the school, 'School is like . . .' The metaphor was used to uncover the types of response to the school which would not be discernible in observation, and would not necessarily emerge in interviews. For this purpose we also used an association list, in which interviewees were asked to give the first thing that came into their head on a series of prompts. These prompts were: school, teacher, pupil, girl, boy, child, young person, British/Finnish, researcher.

The practice of reading in a multilayered sense, either within one section of data or across data, is exemplified by contradictions discerned in our metaphor analysis. The metaphors surprised us with their negativity; references to total institutions (such as prison) were common. Yet the 'prompt' school produced more descriptive terms, associations such as 'teacher', 'friends', 'learning', 'homework', although there were also negative references such as 'a boring place'. In the interviews students often described their own school using many positive terms. "Metaphors translate, invent and betray . . . [they] both conceal and reveal . . . clarify and confuse" (Gordon & Lahelma, 1996, p. 303). Metaphors tell one story whilst neglecting others. This was evident when we asked students to explain metaphors in interviews. A high achieving student explained her 'prison' metaphor:

> Well, somehow, because you can't get away, if you don't want to be here, you can't get away, you just have to be at school. And then you have to be good, or at least I got that, I want to do well at school, so I'm forced to be here (FSH).

We concluded that the metaphors referred to the physical school in general, and to restrictive time-space paths in particular (Gordon, Lahelma & Tolonen, 1995; Gordon, Holland & Lahelma forthcoming). It was evident that different data could produce different kinds of result. Because of this, we turned our attention to the gaze of the researcher in our methodological exploration.

THE RESEARCHER'S GAZE

Researchers in the social sciences today confront a double crisis, of representation and legitimation associated with the interpretive, linguistic and rhetorical turns in social theory (Denzin & Lincoln, 1994). The representational crisis questions whether qualitative researchers can capture lived experience in their texts, or whether they create it. The legitimation crisis involves a serious rethinking of terms like validity, generalizability and reliability, in the poststructural moment. In this moment the concept of the aloof researcher has been abandoned and the qualitative researcher is recognized as historically positioned, locally situated and a very human observer. The demands on the ethnographer are to engage with critiques of the activity, and in reflexive examination of practice. It may also be important not to lose entirely the history of ethnographic and other qualitative work, since there may be insights to be gained from a consideration of the shifting epistemological stances and paradigms which have played across the field, and the continuities and discontinuities in method, methodology and practice.

Interpretation starts before we, as ethnographers, go to the school; it continues when we decide what to write down, and when we read and reread the notes and relate them to other kinds of data that we have collected. As the Finnish researchers conducted their ethnographic research cooperatively, we have all met and observed the same people and thus undertaken joint interpretation. For example, we discussed Manu: Manu's spatial praxis veers towards the creative and active, involving shaping and reshaping. He is habitually late, he frequently walks about during lessons (when the students are obliged to remain seated), he hides under desks, he has conversations with people through the window. Often when he acts in a manner required of a professional student, he dramatizes and ritualizes his own activities and, implicitly, the official praxis in the classroom. When Manu sits on the table in the home economics class and says "I've got my periods – the table is getting wet," this episode is open to a myriad interpretations: it could for example be resistance, but Manu is also engaging in a process of challenging and controlling girls in the group (Gordon, Holland & Lahelma forthcoming). There are occasions when researchers have to accept that they cannot arrive at a conclusive interpretation.

Education researchers who have spent time in schools and classrooms are more than likely to have noted how, in the course of lessons, teachers often turn their attention to those students whom they consider to talk or move inappropriately. By turning our reflective gaze towards our own research practices we have argued that the researcher's gaze is often drawn in similar directions (Gordon, Holland, Lahelma & Tolonen, 1997). We analyzed our notes from a sample of lessons and noticed that many of the observations were on boys doing and talking. Observations on girls focused more on being and looking. Altogether, our data include many more notes on boys' than on girls' visible and audible activity.

The visible activity of boys that we registered was often described as control, abuse, embodied, spatial, competitive, sharing, dramatizing, criticizing. When we talk about girls, we use the words share, observe, communicate quietly, act competently, turn inwards, smile, blush. Sometimes we decided to turn our gaze specifically onto girls, as Tuula did during one science lesson. Her observations from this lesson included girls writing notes, sending materials to each other, having already completed the experiment, not making mistakes, not experimenting (innovatively), having cleared up things ages ago, sitting, waiting and looking at boys doing, and making a show with different colours.

When audible interaction catches the researcher's attention, it is very often loud boys who draw it. These actions may be part of the teaching of the official school, but also in opposition to it – and still be allowed by the teachers. Girls' audible action is most often interrupted, and is usually redirected to the business of the official school. Stillness and silence are not often mentioned in our notes. Some boys may sense this focus on action and test it by going to sharpen their pen in a quiet lesson, looking at the researcher to see if she's noting it (Gordon, Holland, Lahelma & Tolonen, 1997).

There is, however, another tendency in our gaze. Ethnographers in general tend to be more oriented towards differences than towards commonalities (Spindler & Spindler, 1982). A girl's act is more easily registered when it is not expected within the cultural conventions of femininity. As feminist researchers we have tried to be sensitive to the obvious stereotypes, for example the 'quiet girls'. We have noted girls' disputes, resistance or disruptive action more easily than that of boys. Since we still have far fewer notes on these incidents enacted by girls, we can say that there are gendered patterns. It is important to add, however, that there are differences between different girls and boys, as well as between different groups.

GIRLS TOGETHER

Although girls' interaction with each other within the official as well as informal school did not reach our notes as often as boys' interaction, our data are rich with examples of friendly cooperation among girls, helping each other while the teacher spends time with boys, and supporting each other when boys pick on them. Girls might also reveal deep emotional bonds with each other:

> Aura is [my best friend in the class]. I cling onto her, because she has been with me since the first grade. And she's the only one from our class who has come [to this school]. So because of that I cling onto her, and she sometimes clings onto me (FSH).

Girls' cooperation with each other was often silent and invisible to the teacher, as it was to us. Valerie Hey, in her study of girls' friendships, was interested in notes that girls wrote to each other. When she asked girls why they did that, they explained that "if we say things out loud people will hear that we are having arguments and start picking on us" (Hey, 1997, p. 56). Girls may have had such harsh experiences that some of them fear boys; one girl explained that she and her friend burst out crying when they saw that their new class was full of boisterous boys. We observed examples of girls supporting each other when teased by boys.

> Two girls are the 'stars' in the class, Mary and Lisa. An argument develops between Peter and Mary, he says she is looking glum (probably because she did not come top in the Science test). She fights back "I might look stupid, but I don't get bad marks," and Lisa, who did come top on this occasion supports her "Thanks Peter, when you think you've done badly, you don't need that, thank you very much." The two girls have an interesting friendship, based on competition and care; at times they are very tactile, at others quite frosty. (ObsI.)

It is easier for Mary to fight back when she can (usually) rely on Lisa's support.

In another teaching group where there are both noisy and quiet girls and boys, Henna acts like boy soloists, but this position is not as easily available for her as it is for boys. Our observations indicate that teachers control her; when several students may be giggling and shouting out 'silly' answers, teachers may reprimand only Henna. Henna's voice is loud and 'girlish', and she sometimes annoys teachers. She challenges boys when they mess around, telling them to stop talking or throwing things, and appealing to the teacher. But direct challenge may be difficult; her noisy complaints are usually supported by notions of professional pupilhood and school rules. In the interview Henna describes her class by referring to divisions between girls and boys. The interviewer asks her to say more:

> *Henna*: Well I don't know if they're OK, at least with Manu and Pete and them. They always interfere with other people's business, at least if we have said something . . . you get lots of comments. – It's as if you weren't allowed to say anything at all. You should just be quiet and not be yourself.

> *Elina*: Do you think it's particularly you who get comments?

> *Henna*: That's it – not the others so much. You know, every time I open my mouth, I always get comments. As soon as I say something, I can hear 'jäkä, mäkä, mäkä' behind me. It's always like that: 'quiet!'

Elina: How do you feel about that?

Henna: Sometimes it's really frustrating. I get really irritated when they always say something when I talk. You always get commented on, you know, 'she's wearing childish clothes', "she's a bit childish" and so on (FSH).

The tense moments described here reveal the vulnerability experienced by girls. The examples indicate that girls have learned to watch out for the boys who are successful in the informal school. Some girls have gained a reputation of being 'hard'; they have crossed the threshold of 'ordinary girlhood', and have more space and leeway than 'ordinary' girls; for the 'hard girls' surveillance is not suspended to the same extent:

Like Netta and the others, they're so hard anyway, so they don't do it so much to them, but if someone sort of normal, a bit more quiet girl starts to act like that then it would happen (FSH).

But there are many ways in which gender relations are played out. Other enactments by girls, in relation to boys, are withdrawal, conformity and self-discipline, or challenge, resistance and physical self-defence, and include joining in as audience or sometimes as participants. The following extract from a lesson of home economics demonstrates Henna, Marjaana and Tiina in a hilarious game with Manu and Eetu, boys from another group who try to invade the classroom. At the same time the girls challenge another boy, Lasse, from their own group.

A knock on the door. Juuso opens it. Manu and Eetu stand at the door. They whisper asking for [oat biscuits]. Some messing around at the door with them. Teacher makes some comments. Henna, Marjaana and Tiina go to the door, they giggle, play-act with Manu and Eetu, struggling to close the door. They chase Manu and Eetu out, close the door, the boys knock again, the girls knock back from the inside; laughter. Lasse comes to the door and says to the girls, 'Babies'. Girls challenge back: 'Who's a baby, Lasse'. The teacher comes, opens the door to the boys and says they are not to knock any more, not once. (ObsH)

In the extract above the teacher stopped the game and the girls returned to the theme of the lesson. Most of the time girls participate in the official agenda, often helping each other.

In a maths lesson in which the class is unusually quiet, having had the experience of being kept silent and doing nothing but wait for a culprit to admit to a misdemeanour in the previous English lesson, two girls are talking softly, they help each other with the work. In the following lesson, a very active, participative History lesson, with many different teaching methods used and a lot of activity focused on the aims of the lesson rather than disruption, the shared working support continued between these girls. When April read aloud and came across difficult words, Kathy would whisper them to her; she also helped her with her spelling when there were writing tasks (ObsL).

The following notes are from a lesson with the group which includes quiet girls. The teacher interacts mainly with boys, joking with them, giving them commands, appealing to them. Notes on girls include the following observations:

> Nelli and Taru exchange a few words.
>
> Nelli answers.
>
> Milla sits. She fiddles with her pen.
>
> Milla smiles.
>
> Heini helps Soila.
>
> Taru asks something from Milla, about the exercise, I assume. Hille asks for Milla's advice. Milla says she has calculated it in her head. 'You won't be able to do it anyway'.
>
> Milla seems to be ready. She looks toward the window. She fingers her nose. She looks at her exercise book. She rubs something out. Taru turns to her and asks something.
>
> Nelli asks Milla if she is done. Nelli asks how she got the answer. Milla explains.
>
> Milla is knocking her foot on the floor.
>
> Milla asks something from the teacher. Hille asks for an exercise book (ObsH).

This extract indicates that girls help each other whilst the teacher is occupied with noisy boys. Everybody turns to Milla, a quiet, able girl. She is helpful, and girls seem to get more indivdual help from her than from the teacher. But internal dynamics among girls are also indicated; although the girls support each other, Milla will not help Hille, a quiet working-class girl whose school achievement is rather limited. Milla responds in a belittling way to Hille's request for help.

The above examples refer to the official school, but supportive interaction among girls is often embodied and spatial, too. Girls make contact across space in various ways, extending spatial restrictions they encounter. They move their desks together, hold hands, comb each other's hair, lean on each other, hug, write notes to each other, quietly make funny faces egging each other on, wave fingers, clap hands, hum tunes, whisper, giggle, talk etc. Their interaction at times is very embodied:

> Pinja leans on Heini, with her head on her shoulder. Ida leans on Noora. And Jatta leans her elbow on Ida's shoulder (ObsH).

Mutual cooperation among girls was frequently observed by us, yet in the interviews, although girls do talk about their own friends, divisions and conflicts among girls are often emphasized. It was striking to hear girls describing how they did not get on, or how groups of girls did not get on with other groups of girls. Therese gives an example of trouble and strife among girls in her class.

> All I remember is us arguing and shouting at each other and then talking behind our backs and stuff, like. Me and Sandra, we didn't do anything and, erm, we sometimes give each other funny looks, yeah just, just for the hell of it, yeah, and my friend took it really seriously and all this stuff and we were trying to say we weren't saying it about her, we were, because erm, we try and make each other laugh, yeah, me and Sandra . . . and erm, that's what we were doing and she took it seriously and so she turned . . . Christine against us, and we didn't know why, we hadn't done anything. Then she started calling us hypocrites and all this stuff, and we were like, oh fine, if that's how you want to play it, then fine, so we just took shots at them as well (FSL).

Girls interact in supportive ways, although there are times when the internal dynamics erupt, often in ways which the girls cannot quite work out; 'we didn't do anything', as is suggested above. Raisa said that she liked her class but there were disputes among girls every now and then.

> Well, normally they come even from small things. For example, I reserved a desk for Marianne. But then Riikka came and sat on it. And then it turned into a small dispute. If one says kind of 'shut up', then another says 'you shut up yourself' and everything. So that a big fight can start from a small thing (FSH).

Some of the conflicts amongst girls have to do with social divisions, such as ethnicity, social class and sexuality. For example immigrant girls might be excluded. Nicole's fieldnotes describe Betty crying during one lesson.

> I was sitting next to Betty and asked whether she had painted the Eiffel Tower because she had been to France or would like to visit Paris. Her response was that she would like to return to Kenya to feel free and be with her friends. This provoked sobbing tears. [In this instance both girls and boys actually came to ask her what was wrong.] Betty revealed that she felt lonely and isolated in Britain. She had not made any friends outside school and felt that the girls in the classroom ignored her. Betty attributed this hostility to her blackness because in Kenya she was a popular girl.

The above extracts suggest that there are tensions between our observations and what girls said in interviews. To an extent these tensions are also visible in feminist research: on the one hand girls seem very supportive of each other, and on the other they seem capable of causing each other considerable hurt. We address these tensions next.

CONTRADICTIONS

Valerie Hey (1996), in her analysis of girls' friendships, found them to be weighted with difficulties. She presents an "ambivalent account of girls located in economies of friendships as sites of power and powerlessness" (p. 19). She notes that there has been little research on same-sex relations among girls; these have been "variously overlooked, overromanticized, overpoliticized and oversimplified" (p. 6). Hey criticizes masculinist theories, but also asocial accounts of girls' friendships, as well as "girl-friendly" theories for the "tendency to idealize girl – girl relations on the assumption that they are male-free" (p. 15). Her own data, in contrast, are "ambiguous, representing fragmented and contradictory feminine subjects with contextually variant assertive as well as diminished voices" (p. 9). Valerie Hey suggests that in order to write about the positive aspects of girls' relations it is necessary also to write about divisions and difference among them.

In her ethnographic account and analysis Valerie Hey tends to concentrate on the negative aspects of interaction among girls. She writes about bitching, arguments, exclusionary practices and girls' projections of their own ambivalences and contradictory feelings onto other girls. The 'incentive for 'othering' is, she suggests, enormous in the margins of the school. Like Valerie Walkerdine & Helen Lucey (1989), she refers to the need girls have not to stand out and to be 'normal', so that any badness in them is placed on other girls.

We concentrate on Valerie Hey's study, because there is little else in the literature to help us to examine our contradictory findings. Yet although her analysis is enlightening, we are not entirely convinced by it, seeing her as placing too much emphasis on trouble and strife among girls. Her concern is not with supportive relationships among girls and, to make her point, she may overemphasize the strife. Her argument is forceful, and takes full account of the complexities of girls' lives, and yet we want to take issue with her analysis.

But we must ask ourselves, what is the source of our unease with her emphasis on exclusionary practices among girls? Are we romanticizing? Do we want to idealize relations among girls? To explore this, it is useful to start with the personal. Valerie Hey refers to her own memories of being a working-class girl who wanted to do well at school which made her different from other girls with whom she had a difficult relationship. She was accused of having "become a snob": 'I still reflect upon this incident and its aftermath. Had I betrayed my friends or had they betrayed me? The politics and pains of my class aspiration were played out in those relationships, by gendered cultural means and with specific cultural effects' (p. 31).

These memories are intertwined in her analysis, both consciously and unconsciously. There were girls in her study who were similar to the ones she had grown "with and against" (p. 89). She discusses methodological implications of her painful experiences:

"My intention in claiming resonances between an ethnographic text and aspects of my biography is not to stake a privileged claim on truth. Rather it is to recognize the significant (if immeasurable) effects of personal history. The practice of re/membering has been surprisingly emotional. In getting to know Carol and other schoolgirls I have been continually reminded of resonances from my own girlhood. At a deeper level it is, however, 'difference' that constructs our relation and relationship and my rendition of it" (p. 89).

We conducted memory work on our own schooldays, Tuula and Elina with the researchers in the Helsinki team, and Tuula, Janet and Elina together. We wanted to find the schoolgirl within us, to be as aware of her as possible and to know when we were looking or interpreting through her eyes. We started the memory work before the fieldwork and again before analysis, but also undertook it during our analysis of relations among girls. Valerie Walkerdine (1997) suggests that one's own experiences can, in a sense, validate one's analysis. We have not approached our memory work in this way but have seen it as a methodological device which, although drawing on our own experiences in order to strengthen our understanding, aimed to free us from constraint on what we see and how we interpret what we see through our schoolgirl eyes. It helped us to be reflexive about our own approach and, for example, to problematize our own gaze, as discussed above.

When discussing the contradictions in relationships among girls, we considered our own memories of interaction with other girls and women. Though we have had bad, hurtful, infuriating and painful experiences with women, overall we felt that a great deal of our interaction as girls with girls and as women with women has been supportive and sustaining. Our commitment to collaborative work with women (though we have not excluded men) is one result of those positive experiences. In our cross-cultural, comparative research collaboration has been a real strength – so our experiences have influenced ways in which we practice as researchers.

All this has implications for the ways in which we have tried to solve the contradictions in our data and led to our feeling that positive aspects of girls' relations should be considered, whilst not neglecting divisions and the ways in which these were played out. One aspect of Valerie Hey's work provided a key to why we felt that collaboration among girls was more prevalent and more important than she suggests. She argues that girls played out their differences in particular spaces: "the backstage of classrooms; the periphery of schools; the playing fields: the local off-site amenities: the cemetery; the local cafe; the shops; the canteen, and the school yard" (p. 45). Our observations and participation were confined to the school. Our interest was also very broad – we wanted to explore processes of differentiation in everyday life at school, so our data went into less depth on girls' relationships.

The varied spatial settings Valerie Hey talks of are, however, relevant in other ways. Such marginal spaces were safer for girls to work out their differences, without control or interference from others. These were also spaces from which girls could escape when they wished. School space, however, cannot be left at will and social situations in school have to be handled in some way. Whilst students have to learn to be alone in the crowd, the crowd is safer if one has friends. The research team in Helsinki experienced the buzz and the nervous excitement of starting school, when students were milling around, trying to hang on to those they knew, to make the acquaintance of those they did not, but whose clothing, style and bodily comportment signified potential for sharing, and to avoid those whose bodily insignia indicated likely indifference or possible threat. Jenny Shaw (1996) notes that having or finding a friend reduces the chaos of the school for students. School seems a safer place if one has friends, or at least groups that one can belong too, or on the edges of which one can move.

Our observations indicated that girls were subject to control by boys and their movement and voice were more likely to catch the teacher's gaze. Interaction among girls was important and helped them in the process of making spaces. Girls were usually (though not always) less salient in the classroom than boys; past experience of controlling actions by teachers and restrictive behaviour by boys contributed towards a more quiet demeanour among them. It is easier for girls to find a place in the classroom which grants them an acting position, if they have established patterns of interaction amongst themselves. Nelli's time at school is spent in two different teaching groups. In one of them the boys are boisterous and noisy, and the girls are quiet. In the other, Nelli explains, many students talk, and girls join in. In the first, as she does not know anyone well, she remains relatively quiet. But as she comments:

> as you talk with others . . . then you shout out the answers to the teacher as well at the same time . . . if you know someone . . . but if you are alone in the class then you end up saying nothing and you don't raise your hand (FSH).

Nelli refers to 'shouting out' answers. Our observations suggest that Nelli is not in the habit of 'shouting out'; she tends to put her hand up and wait to be given her turn to speak – although she will shout out occasionally if she needs help and the teacher does not notice her. But her choice of the phrase 'shouting out' indicates that Nelli has a sense of having a voice through supportive interaction among girls.

Sonja who is in the same two teaching groups as Nelli also gets frustrated by the boys. Sonja is a self-confessed 'bod' who does well at school and hopes for space for concentration, particularly when writing essays. She argues that girls' quietness is partly a consequence of the control exerted by boys:

> Girls can't shout so much. If they could shout like boys, then there would be pointed fingers, you know, that's not feminine at all (FSH).

Nelli does challenge boys; though she talks about being quiet during the lessons, she is also assertive and 'answers back'. She concurs with this in the interview, but adds that when you do talk back, you must harden yourself to hearing comments for about a fortnight afterwards.

It seems evident from our observations and from interviews that girls need each other's support and often get it. Although girls refer to boys controlling them or at least cramping them by being so boisterous, interaction among girls can redress the balance. Girls are important for each other; if they are marginalized in the informal school, they can crawl from the margins through mutual support.

When we place the informal school in its entwined relation to the official school, girls' joint support seems crucial in helping them to make spaces. We focus on a whole range of interactional patterns among girls. Valerie Hey also wants to redress the balance of research on girls and schooling; girls' interpersonal relations and ways in which they are positioned and position themselves in schools do not create a girlfriendly nirvana, nor oppositional heroines (1996, p. 126), and we agree. The marginal positionings that girls inhabit, the 'borderlands' which they visit and revisit, become more habitable places from which they can also escape when girls interact to develop mutual strength. But such positionings are also fraught and ambivalent; whilst making connection, they also talk about splits and divisions which are located in the axes of difference among girls. Considerable emotion and desire is embedded in girls' interaction with each other, as Valerie Hey (1996) and Jane Kenway & Jill Blackmore (1996) suggest. Ida talks about tensions among girls in her form, but she adds:

> Sometimes we have unbelievable fun, you know, sometimes we're numb with laughter in the dining room, when we remember some stupid thing (FSH).

Girls place high expectations on their friendships, and so arguments among them can be taken more seriously than those amongst boys. But we may also have expectations that girls will not argue, when friendship does not of course preclude arguments. Marjaana explained that she and Henna often argued, but that Henna was a really good friend. Cultural representations of female bitchiness are influential in the overemphasis on trouble and strife and underemphasis on mutual support. Boys in our study tended to say that they moved in crowds and that they were all mates together, but their jostling to construct and maintain hierarchies among themselves were a much more visible and constant part of everyday life at schools than the construction of hierarchies among girls.

We have used our data in this analysis, but we suggest that the problem we set out to explore – to explain contradictions between our observations and the interview data – cannot be solved through data alone. We had tried to trace our steps and to illustrate how we have arrived at our analysis, but we were unable to do so without interpretation. Interpretation, the process of trying to understand the meaning of results, is an act in which researchers employ a whole range of tools, some more, some less consciously. Our own personal experience influences our interpretations, as does our political stance.

In that process we disengage ourselves from naturalistic ethnographic research where scientific methods are applied and objectivist accounts are produced, as Leslie Roman (1993) suggests. But, like Roman, we also disengage ourselves from subjectivist research, where the ethnographer comes to understand cultural process and meaning-making of people in the field, while producing accounts which are value-directed, so that all accounts are equally acceptable. We do that by avoiding dematerializing our account (cf. McRobbie, 1997; Morley, 1997) – we do not ignore power relations that form the context of processes of differentation and marginalization in schools.

Whilst we have been making methodological and analytical choices, the interpretive element means that our choices are always cultural and political too. We have tried to demonstrate the process of interpreting contradictions in our data, and to explain why we emphasize the importance of mutual support among girls – whilst not neglecting painful processes of differentiation among them – but we concur with Valerie Hey when she suggests that all interpretations are "always provisional and partial" (p. 144).

FRIENDS AND FOES

Our search for reconciling contradictions in our data also stems from a broader approach we had undertaken before we entered the schools to conduct our research. We wanted to make a break with studies that have emphasized conflict and resistance in schools. We did not want to concentrate merely on repressive aspects of schooling. We had resolved to look for stories of cooperation as well as stories of conflict. Multilayered processes are taking place in schools – we wanted to gaze at as many of these as possible. We developed methodological and analytical approaches which would allow us to retain the multilayered nature of schools. There have been times when we have suggested that a particular incident is open to a range of interpretations, and we have not been able to say which interpretion is 'correct'; we have suggested that researchers have to accept that the practices and processes they observe can defy analysis and that we ought to accept that. But at other times, as here, we have wanted to read and reread our data, analyze it and reanalyze it, and to step up the interpretive process whilst trying to be as conscious as possible of the social, cultural, political and personal stakes we draw on when making our interpretation. As feminists analyzing interaction among girls we felt that this was the point where we wanted to move beyond stating that there are a range of possible interpretations.

We have suggested that the support that girls give to each other in the official school is very important to them. If there are boisterous boys in the classroom who draw the attention and energy of the teacher, girls can help each other. Moreover, their cooperation makes the classroom a more pleasant place for them. In the physical school cooperation helps them to broaden the space that they are using, facilitates their use of voice and gives their bodies more scope. In the informal school friendships make schooldays more fun – though particularly in the informal spaces girls also play out their differences and engage in exclusionary practices, marginalizing some girls or some type of femininities. Trouble and strife on the one hand, and cooperation and mutual support on the other, are contextual practices in which being 'foes' in one spatial and social situation does not preclude being friends in another.

Restructuring has brought the market into schools in Britain and in Finland (though more so in the former). At the same time resources have not been adequate. Competitition between schools and within schools is encouraged whilst teaching groups have increased in size. Teachers have been under a great deal of pressure in their work as New Right policies have tightened everyday life in school. There is less time for dealing with contradictions and tensions, and less opportunity to make space for or to encourage supportive relations among students. We have argued that such support is important for girls – but this applies to all students.

NOTES

1. ObsL/ObsH refer to observation notes from London/Helsinki schools. FSL/FSH refer to the interview of a female student in London/Helsinki schools.
2. Our thanks are due to Kay Parkinson for suggesting this metaphor.
3. Consider the horror caused by 'unacceptable' steps and movements in ballroom dancing in the film *Strictly Ballroom*.
4. Helskinki: Pirkko Hynninen, Tuija Metso, Tarja Palmu, Tarja Tolonen; London: Nicole Vitellone and Kay Parkinson.
5. From the Finnish schools we have 96 student interviews (51 girls and 45 boys), 44 teacher interviews (34 male and ten female) and four support staff interviews (one male, three female). In the London schools we have 71 student interviews (39 girls and 32 boys), 16 teacher interviews (ten male and six female) and three support staff interviews (one male, two female).

REFERENCES

Burr, V. (1995). *An Introduction to Social Constructionism.* London: Routledge.

Delamont, S., & Atkinson, P. (1995). *Fighting Familiarity: Essays on Education and Ethnography.* Cresskill, NJ: Hampton Press.

Denzin, N., & Lincoln, Y. (Eds) (1994). *Handbook of Qualitative Research.* London: Sage.

Giddens, A. (1985). Time, Space and Regionalisation. In: D. Gregory & J. Urry (Eds), *Social Relations and Spatial Structures.* London: Macmillan.

Gordon, T., & Lahelma, E. (1996). 'School is like an Ants' Nest' – Spatiality and Embodiment in Schools. *Gender and Education, 8*(3), 301–310.

Gordon, T., Holland, J., Lahelma, E., & Tolonen, T. (1995). Koulu on Kuin – Metaforat fysisen koulun analysoinnin välineinä. *Nuorisotutkimus-lehti 3,* 3–12.

Gordon, T., Holland, J., Lahelma, E., & Tolonen, T. (1997). Hidden from Gaze: Problematising Action in the Classroom. Paper presented at the British Sociological Association Annual Conference.

Gordon, T., Holland, J., & Lahelma, E. Forthcoming. *Making Spaces: Citizenship and Difference in Schools.* London: Macmillan.

Hey, V. (1997). *The Company She Keeps. An Ethnography of Girls' Friendships.* Buckingham and Philadelphia: Open University Press,

Kenway, J., & Blackmore, J. (1996). Pleasure and Pain: Beyond Feminist Authoritarianism and Therapy in the Curriculum. In: P. F. Murphy & C. V. Gipps (Eds), *Equity in the Classroom,* London and Washington: Falmer Press.

McRobbie, A. (1996). Looking Back at the New Times and its Critics. In: C. Kuan-Hsing & D. Morley (Eds), *Stuart Hall: Critical Dialogues in Cultural Studies.* London: Routledge.

McRobbie, A. (1997). The Es and the Anti-Es: New Questions for Feminism and Cultural Studies. In: M. Ferguson & P. Golding (Eds), *The Cultural Studies in Question.* London: Sage.

Molloy, M. (1995). Imagining (the) Difference: Gender, Ethnicity and Metaphors of Nation. *Feminist Review, 52,* 94–112.

Morley, D. (1997). Theoretical Orthodoxies: Textualism, Constructivism and the New Ethnography in Cultural Studies. In: M. Ferguson & P. Golding. *Cultural Studies in Question,* London: Sage.

Roman, L. (1993). Double Exposure: The Politics of Feminist Materialist Ethnography. *Educational Theory, 43*(3), 279–308.

Shaw, J. (1996). *Education, Gender and Anxiety.* London: Taylor and Francis,

Spindler, G., & Spindler, L. (1982). Roger Harkes and Schönhausen: From the Familiar to the Strange and Back. In: G. Spindler (Ed.), *Doing the Ethnography of Schooling. Educational Anthropology in Action.* New York: Holt, Rinehart and Winston,

Walkerdine, V., & Lucey, H. (1989). *Democracy in the Kitchen: Regulating Mothers and Socialising Daughters.* London: Virago.

Walkerdine, V. (1997). *Daddy's Girl: Young girls and Popular Culture.* London: Routledge.

UNDERSTANDING HETEROSEXUALITIES: MASCULINITIES, EMBODIMENT AND SCHOOLING

Mary Jane Kehily

INTRODUCTION

This chapter considers the relationship between heterosexuality and masculinities in educational establishments. The focus is upon the ways in which young men in school constitute and consolidate heterosexual masculine identities. Through an analysis of interviews with young men in school, the chapter suggests that school processes produce sites for the enactment of heterosexual masculinities. Furthermore, these enactments demonstrate both the normative power of heterosexuality and the fragility of sex/gender categories. In the lives of young men in school, heterosexuality is understood as a *practice* involving a set of social performances in relation to young women and other males. Among the young men I spoke with there was little understanding of heterosexuality as an institutional arrangement for the support and maintenance of a particular sex – gender order. Rather, heterosexual relations were viewed as a way of demonstrating a particular masculinity that could be utilized to command respect and confer status on some males while deriding others. A theme of the chapter explores issues of embodiment as expressed by the young men. In these exchanges there is an emphasis on the physicality of the body,

Genders and Sexualities in Educational Ethnography, Volume 3, pages 27–40.
Copyright © 2000 by Elsevier Science Inc.
All rights of reproduction in any form reserved.
ISBN: 0-7623-0738-2

often articulated in terms of activity and performance, where the physical sense of maleness is constantly reiterated as 'doing' heterosexuality.

METHODOLOGY

This chapter draws upon material from my doctoral research: an ethnographic school-based study which aims to explore issues of sexual learning in relation to young people. The study looks at two key areas in the field of sexuality and schooling; first, the construction of sexual identities within pupil cultures and second, how school processes shape the domain of sexuality through the curriculum and institutional practices. The research uses a wide range of ethnographic methods including participant observation, group-work discussion and semi-structured interviews. The fieldwork for this study was carried out over a total of two years, beginning in 1995 and continuing in the academic year 1996–97. During this time approximately 180 interviews and discussions were carried out with teachers and students.

Ethnographic research was conducted in two secondary schools in different parts of the U.K. Both schools were open access secondary schools for boys and girls aged 11–16 and both served a largely working-class catchment area. Oakwood School was non-denominational and racially mixed with many pupils from south Asian and African-Caribbean backgrounds. Clarke School, however, was a Church of England school and the school population was mainly white. I supplemented the ethnographic fieldwork with focus group discussions at an all-boys secondary school in a large city in the south-east of England. This school was accessed through personal contact with a senior teacher in the school who described the student population to me as 'very mixed' with a high percentage of students from African-Caribbean, mixed heritage and south Asian backgrounds. The material upon which this chapter is based is drawn from group-work discussions with young males at the school in the south-east of England. The boys, aged between 14 and 15, were selected for me by their teacher as representing a cross-section of the student population in Year 10. Though the composition of the group changed during the course of the school term, their social positioning can be described as working class and racially mixed, including boys who identified as African-Caribbean, south Asian, black British and white British. The one-hour discussions I conducted with them can be described as open-ended; although I expressed an interest in talking to them about sex education, I was happy to let the conversation stray into a range of themes and issues that emerged from them as a group. Generally, my approach to discussions was to intervene as little as possible in the flow of dialogue between the young men, only asking questions as a response to their interactions and occasionally seeking clarification and development of themes.

The ways in which students spoke to me about sex involved me in many encounters which provided an opportunity to reflect on the relationship between sexualities and the social context of the school. Students utilized sexuality in a variety of ways in their interactions with teachers and with each other. These sexualized exchanges provide an insight into the sexist and homophobic practices of pupils to suggest ways in which sexual power is played out in school (Lees, 1993; Mac & Ghaill, 1991). Anthony Easthope (1990) suggests that talking dirty and particularly the sharing of 'dirty jokes' between men is an attempt to 'master' women through discourse (1990, p. 126). The comments of young men in my study could be viewed as an attempt to invoke a form of 'mastery' capable of placing me in a subordinate position to them through the utilization of a deliberately transgressive sexualized style. As a former teacher of sex education, however, I felt familiar with the ways in which young men talked about sexual issues and was not shocked or offended by 'dirty talk'. Furthermore, I came to understand the 'talking dirty' discourse of young men as a preliminary stage in establishing field relationships. Through jokes, sexualized banter and daring questions, young men collectively contributed to a style of bawdy excess that 'tested' the dynamics of social encounters with others. As a researcher interested in sexualities and schooling, pupil peer groups provided me with a rich and valuable insight into the ways in which young people spoke about and articulated issues of sexuality. I came to understand these exchanges as constitutive of informal sexual cultures within the school. As field relations developed, the sex-talk of students provided me with access into their perspectives and also opened up a space for me where sexual themes could be pursued in interviews and groupwork discussions.

APPROACHES TO THE BODY

The idea that the body is an important site for the exercise of power can be located within a Foucaultian framework where the rise of capitalism can be seen to create a new domain of political life, referred to by Foucault as 'bio-power' (1978, p. 140). Here power is conceptualized as de-centralized and productive of social relations in commonplace encounters and exchanges. From this perspective the politics of the body plays an important part in disciplining individual bodies and regulating collective bodies such as populations or specific social groups. For Foucault the body is discursively constructed, realized in the play of power relations and specifically targeted in the domain of the sexual. Foucault sees sex as a political issue, crucial to the emergence and deployment of bio-power,

> It [sex] was at the pivot of the two axes along which developed the entire political technology of life. On the one hand it was tied to the disciplines of the body: the harnessing, intensification and distribution of forces, the adjustment and economy of energies. On the other hand, it was applied to the regulation of populations, through all the far-reaching effects of its activity (Foucault, 1978, p.145).

For Foucault, disciplining the body at the level of the individual has a historical trajectory that can be traced to the Christian pastoral tradition of the seventeenth century. Christian spirituality encouraged individuals to speak their desires in order to control them. The process of transforming desire into discourse in a religious schema has the effect of purifying the mind and the body by expelling worldly desire and turning back to God. This spiritual experience produced, for individuals, "a physical effect of feeling in one's body the pangs of temptation and the love that resists it" (1978, p. 23). Foucault points to the links with the sexual libertine literature of the nineteenth century such as *My Secret Garden* and the writings of de Sade where sexual activities and erotic attachments are described and documented in episodic detail. One way of understanding this 'tell all' experience is to view it in terms of internal relations or psychic structures whereby the Other is produced within the Self. From a psychoanalytic perspective this has the effect of heightening desire by producing the 'forbidden' and, simultaneously, heightening anxiety in the constant struggle to expel the Other from within. It is possible to view the homophobia of young men in school as part of this dynamic. Similarly, the desire/repulsion expressed by girls in relation to sexual activity can also be seen as an internal dynamic, variously played out in social arenas. In these examples the desire for/fear of relationship is enacted within the peer group and plays a part in the structuring of heterosexual hierarchies in school (Kehily & Nayak, 1997). In peer group interactions, individuals are active in the control and regulation of their own bodies within a broader context of control and regulation.

Colette Guillaumin's (1993) study suggests that "the body is the prime indicator of sex" (1993, p. 40) where external reproductive organs are ascribed a set of material and symbolic meanings elaborated in the construction of sexual difference. This separation of the sexes at the level of the body is duplicated by a material social relationship involving the sociosexual division of labour and the distribution of power. Guillaumin indicates that the sexing of the body in society is a long-term project involving work at different levels: physical and mental labour; direct and indirect interventions; the exercise of gender-specific social practices and competences. Bodies are constructed in societal contexts where ways of being *in/with your body* have material effects:

> Restricting one's body or extending it and amplifying it are acts of rapport with the world, a felt vision of things (Guillaumin, 1993, p. 47).

For Guillaumin, the materiality of the body plays a part in the production of gender inequalities which can be seen in the different ways in which boys and girls play, use space and engage in bodily contact. Central to the construction of the sexed body is, in Guillaumin's terms the 'body-for-others', ways of relating to others in terms of physical proximity, which is learned by both sexes but experienced differently. Bodily contact among males in combat and play introduces notions of solidarity, co-operation and control of public space. However, for girls the body-for-others is constructed in the private domestic sphere where the female body is both closed in on itself and freely accessible. From this perspective the materiality of the body is constitutive and productive of gender inequalities in ways that are learned, experienced and lived.

Bob Connell's (1995) study of masculinities is also concerned with the ways in which gender is understood and interpreted in relation to the body. He suggests that the physicality of the body is central to the cultural interpretation of gender. In Connell's analysis, as in Guillaumin's, the materiality of the body is important to individuals and to societal arrangements and can be seen to make a difference to the ways in which gender is learned and lived. For Connell, masculinity can be defined within a system of gender relations as,

> simultaneously a place in gender relations, the practices through which men and women engage that place in gender, and the effects of these practices in bodily experience, personality and culture (Connell, 1995, p. 71).

Furthermore, Connell indicates that there is a need to assert the *agency* of bodies in social processes (1995, p. 60) in order to understand gender politics as an embodied social politics. Connell uses the term 'body-reflexive practice' to suggest the ways in which bodies can be seen to be located within a complex circuit as both objects and agents of social practice. In this model the body is located within a particular social order where bodily experience is productive of social relations (and socially structured bodily fantasy) which in turn can produce new bodily interactions (1995, pp. 61–62). 'Body-reflexive practice' captures the dynamic interplay of bodily interactions working within societal and institutional constraints and also the sense of agency that suggests that experiences at the level of the body offer possibilities for transgression and change.

BODIES IN SCHOOL: INSTITUTIONS AND THE EMBODIMENT OF MASCULINITIES

Foucault (1978) points to the ways in which schools of the eighteenth century were structured and organized to take into account the sexuality of children,

> the internal discourse of the institution – the one it employed to address itself, and which circulated among those that made it function – was largely based on the assumption that this sexuality existed, that it was precocious, active and ever present (Foucault, 1978, p. 27).

In particular, the sex of schoolboys, Foucault indicates, is constructed as a public problem in and through discursive strategies which encourage the deployment of a range of medical and educational interventions for the control of adolescent boys. In contemporary schooling pupils become the object of disciplinary regimes which aim to control and regulate the (sexed) body as well as the mind. Rules govern the physical use of spaces where pupils move – in classrooms, playground and corridors. The spaces, in their architectural design and layout, also prescribe, to some extent, the type of movement that is possible and desirable. For example, the subject of 'classroom management' taught at teacher education colleges, suggests to student teachers that the learning environment can be shaped in particular ways by the strategic placing of tables, chairs and classroom equipment. Bodies in school can be seen in two ways: collectively as a student body, to be controlled and moved about with ease; individually as bodies to be simultaneously trained and protected. Sexuality, as Foucault points out, can be seen as a feature which structures the ways in which bodies in school are organized and related to.

In secondary schools, through the social processes of schooling, there is an associative link made between the body and sexuality or, to put it another way, the body is seen as a conveyor of sex, and sexuality is seen as an embodied manifestation of the body/sex couplet. The body that emerges in relations of schooling is predominantly heterosexual. Connell (1995) uses the concept of 'hegemonic masculinities' to discuss the relationship between different kinds of masculinity and the ways in which issues of sexuality and embodiment feature in these accounts. Within the framework of hegemonic masculinities there are specific relations of dominance and subordination played out between groups of men. In these interactions heterosexuality assumes a dominant status, while homosexuality acquires a subordinate position in the sex-gender hierarchy. This chapter is concerned to explore the ways in which heterosexuality is constituted and consolidated by young men in school. The ethnographic evidence discussed in this chapter suggests that heterosexuality is constituted in the everyday practices of young men in school. Within the educational institution these practices have the effect of consolidating and validating a particular masculinity.

CONSTITUTING HETEROSEXUALITY: SEX-TALK, MASTURBATION AND PORNOGRAPHY

In the interactions between young men in school, heterosexuality can be seen as a *practice* involving a set of social performances in relation to young women and other males. Among the young men I observed and spoke with there appeared to be little understanding of heterosexuality as an institutional arrangement for the support and maintenance of a particular sex – gender order. Rather, heterosexual relations were viewed by the young men as 'natural' and as a way of demonstrating a particular masculinity that could be exercised to establish a position of privilege within the male peer group. A central theme in the demonstration of an esteemed masculinity is the notion of 'knowing it already' in matters of sexuality. In conversations with young men, I ask them how they learned about sex and was frequently met with responses such as, "I already know about it . . . I taught myself" (Justin); "We already know it, I do anyway" (Blake). In these exchanges certain young men seemed keen to assert, to me and the other boys in the group, that sexual knowledge was located in the self. In these declarations Justin and Blake suggest that achieving knowledge and 'knowing' in the domain of the sexual is acquired through self-activity. Justin elaborated on his sexual learning in the following terms:

> Justin: Well, my Dad, he's hinted at things, he has, yeah. He told me about, well, he never gave me much explanation like, just hints, but they came together, all things, by watching videos, magazines, listening to friends, older brother and just getting to know for myself, you know.

Here, Justin indicates that 'knowing it already' involves the active and protracted process of making sense of multiple sources. Sexual knowledge, far from being easily assumed and embodied within the masculine sense of self, is in fact learned in the piecemeal way described by young women (see Thomson & Scott, 1991). However, within the male peer group, the demonstration of competence and fear of ignorance become familiar tropes in the articulation of a masculinity that is sexually knowing and heterosexually active.

In the male peer group heterosexual activity is valorized and frequently spoken about in terms of conquest and prestige (Wood, 1984). As Christian put it, "All boys claim to be doing it with girls – everyone in the school." However, while males in school may engage in the sexual boast, there is evidence to suggest that the performance is not always believed:

Christian: I listen to what they say but you can't take it seriously, you can't always believe them 'cos they might just be saying that for their mates, to look strong or to make them look bigger.

Here the physical quality of looking 'strong' and 'big' in front of your mates can be seen as a symbolic attempt to display an exaggerated and inflated masculinity, capable of achieving status in the male peer group. Young men reported speaking about sex with each other by recounting details of sexual encounters with girls in terms such as, "I did this with her last night, then I did that." Such interactions indicate the need to maintain a masculine style premised on activity and performance. In these moments the collective structure of the male peer group offers a performative space where heterosexuality and masculinity can be fused and displayed. This space can also be seen to provide a forum for a form of secular confessional where young men disclose details of their sexual encounters with girls. Transforming desire into discourse, in this context, is turned into a boast which seeks validation rather than repentance. As other researchers have documented, sex-talk between males serves many purposes and can be seen to have a range of regulatory effects: policing the boundaries of gender-appropriate behaviour for young men and young women; providing an imaginary ideal of desirable masculinity; bolstering the reputation of particular males; concealing vulner-abilities and producing heterosexual hierarchies (Haywood, 1996; Kehily & Nayak, 1997; Lees, 1993; Wood, 1984).

The following discussion with a group of young men offers an insight into the workings of male peer groups and the links made between masculinity and heterosexual activity. In this context they talk of having a girlfriend and sexual activity as 'natural' and routine:

So, is it uncool to be a virgin?

Christian: Nah, I wouldn't say that

Justin: It depends, when you're younger, no, but when you get to Year 11 man

Christian: But with your friends, if you tell your friends you haven't had sex, my friends anyway, they wouldn't like act up on you, I mean, they try to encourage you to have sex, but . . .

Justin: My friends would be going [in low voice] go on then, go on now then, get there now, have sex, go on then

Christian: Nah, that's no good

Justin: What's no good?

Justin: It's nothing to be scared of.

Matthew: But who wouldn't want to do it?

> Christian: No man, there's a difference man, there's a difference between wanting to do it
> and people telling you to do it. Even if you want to do it from the start man, but when
> people start telling you to do it, it makes you like, you want to give them a challenge, makes
> you want to say no.

In this discussion the young men indicate that the links between masculinities and heterosexual activity are negotiated within friendship groups. Sex with girls is presented as a general aim to be desired and expected, an eroticized 'getting there' for all boys. However, the reflections of Justin and Christian suggest that heterosexual activity is differently appraised. While Justin's friends would urge 'doing it' in a display of hyper-heterosexuality, Christian's friends would offer encouragement without 'acting up on you'. Christian's comments suggest that other modes of behaviour such as acting autonomously and posing a challenge may also be incorporated into the masculine repertoire and may exist as a counterpoint to heterosexual pursuit. In such a *discursive manoeuvre* Christian is able to resist pressure to engage in heterosexual activity, while presenting a masculine sense of self which seeks to maintain his reputation in the peer group. In discussions with young men it is possible to see the male peer group as significant in negotiating meanings attached to sexual activity, where versions of heterosexuality and masculinity are produced locally. In these interactions heterosexuality is invoked as a practice; an endeavour where the 'doing' is valorized by particular styles of sex-talk.

So far the discussions with young men have focused on the practice of heterosexuality in sex-talk where girls become the object of desire and the subject of a contextually constructed male sex-drive. Further discussions with young men, however, suggest that heterosexuality may be constituted in other practices *before* girls are known and spoken about. In the absence of girls, in relationships or in discourse, young men may fantasize about having sex, imagining what it will be like and how they will feel. This fantasy space is usually accompanied by bodily practices such as masturbation. Peter Willmott's (1966) study of adolescent boys in East London reports that masturbation was "common, if not universal" (1966, p. 54) among males between the ages of 12 and 14. His study indicates that masturbation was spoken about in terms of 'discovery' or initiation into the domain of the sexual. One of Willmott's respondents spoke of masturbation in the following terms: "I started at 14. When I first discovered it I went really mad over it. Then after a while it turned me off a bit" (1966, p. 54). Connell's (1995) study also suggests that masturbation plays a part in sexual learning for young males. One of Connell's respondents spoke of enjoying masturbation 'too much'; the intensity of pleasure was such that he felt compelled to stop for fear that it would prevent him from enjoying sex with a woman (1995, p. 104). Through masturbation young men learn about and experience sexual pleasure; however, as Connell points out, they must then discipline their bodies to be heterosexual, where desire is specifically associated with heterosex rather that with auto-eroticism or homo-eroticism.

In the conversations I conducted with young men in school, there was acknowledgement that masturbation was a common practice which met with routine acceptance and denial.

Christian: People have told me you go through a phase where you start, like, wanking yourself off, stuff like that. I don't think it's true because I ain't gone through that phase yet and I don't think I'm gonna

Matthew: You don't know though

Adam: You don't know that

But don't you think that's one of the ways boys learn about sex?

All: [chorusing] Yeah, yeah
Yeah man
Admit it man

You know, through finding sources that make them feel excited and . . .

Christian: Like the computer or something

[laughs]

Matthew: Like my mate right, he's always on the computer saying, 'Look at this' right and he gets mad excited over his computer

Adam: The Internet

Justin: Consolation, that's all it is, cut all that man, how people get like – obviously they don't care about sex. I know this boy who loves porno, guy's mad about, magazines everywhere, videos. He was telling me to watch one and I says, "No, man, I ain't watching that dirtiness," getting excited for no reason, it ain't worth it.

In this exchange the young men agree that masturbation forms part of their sexual learning. Furthermore, they make a link between the bodily practice of masturbating and cultural resources such as pornographic magazines, videos and Internet pages. Justin appears keen to evaluate such forms of excitement for young men as dirty, worthless and second-rate. His comments draw a distinction between the 'real thing', intercourse with a woman, and any simulations of it through pornography-fuelled fantasy. Justin's dominance in this discussion indicates that there is a need to emphasize sex as heterosexual and penetrative, a move that can reclaim heterosexuality as part of a masculine identity. Rachel Thomson's (1997) analysis of young men's accounts of pornography postulates that encounters with pornographic material are "one of the ways in which young men are brought into identification with a collective masculinity" (1997, p. 2). In recounting their engagement with pornographic material, Thomson documents the ways in which individual young men attempt to evade agency by asserting that the magazine/video was obtained or initiated by a friend or a group of mates, or something they just happened to stumble across. Significantly, both Justin and Matthew claim to have a mate who is into porn, while placing their own interests in such material at a distance.

Thomson suggests that for young men, "there is something potentially disempowering or emasculating about porn" (1997, p. 10) as it involves the practice of seeking sex without being desired. Justin's insistence that pornography is 'consolation', a poor substitute for sex with a woman, can be seen as an amplification of Thomson's point. Negotiations with pornography can be viewed as a way of policing the boundaries of acceptable masculinity where sexual pleasure is evaluated in hegemonic terms as male–female intercourse. The recourse to such a hegemonic structure can be seen to have disciplinary effects in the dynamics of masculine hierarchies.

> Christian: I've never bought one, I've found one before, I've had a look, yeah, but that's all, I've never bought one.
>
> [background jokes about Christian and porn]
>
> Justin: This friend of mine, right, he's got loads [of porn magazines] and I've looked at them, nasty man, horrible. All them pictures man. I say put them away, they don't teach you nothing. He might get excited, but me, not exciting or nothing. People who do that, man, they're sad.
>
> Christian: They're sad, very sad.
>
> *[To Justin] And does he (your friend) have a girlfriend?*
>
> Justin: Yeah, he's got a girlfriend, got a nice little woman. I wish I'd seen her, I'd have liked her for myself. But them videos an' stuff are nasty, man, they should ban them videos.
>
> Christian: Some people enjoy them though.
>
> Justin: Yeah, but if a man watches them and they get all excited and then they turn it off and they can't get no ladies, that's why they rape.
>
> Christian: Can't get no ladies that's men rape.

This discussion between Justin and Christian illustrates some of the ways in which young men negotiate and evaluate masculinities in relation to issues of sexuality. Investments in pornography are viewed as a bad thing. While it is barely acceptable to engage with pornographic material, it is totally unacceptable to buy it for oneself. Coveting your friend's girlfriend is acceptable; sharing his sexually stimulating literature is not. Men who use porn are defined as 'sad' and having an attractive girlfriend, it seems, does nothing to redeem them. Justin and Christian mobilize a moral discourse to underline their view that porn is bad and *dangerous*. They suggest that there is a causal relationship between pornography and rape. Men who use porn need women as an outlet for their sexual urges and if they can't get a woman they rape. In male peer groups young men shape the parameters of acceptable and unacceptable sexual practice. Through such interactions young men implicitly produce definitions of desire and deviance that can be utilized as a technique for displaying certain versions of masculinity and deriding others.

CONSOLIDATING HETEROSEXUALITY: RELATIONSHIPS WITH WOMEN

Through engagements with sex-talk, masturbation and pornography, young men *constitute* a version of heterosexuality that is associated with a desirable masculinity. This heterosexual-masculine combination is negotiated collectively by young men in peer group interactions and may be reconfigured in different ways. Further discussions with this group of young men suggest that it is in relationships with women that heterosexuality is actively learned and *consolidated*. In this discussion young men in the group criticized their sex education in school as 'useless' and 'a waste of time'.

> *Would you have liked to have had a proper sex education?*
>
> Justin: Not now, it don't make no difference to me
>
> *Why's that?*
>
> Justin: Umm, because, I taught myself
>
> *So how did you teach yourself?*
>
> [muted laughter]
>
> Justin: By getting a girlfriend
>
> Christian: And you explore her
>
> [mumbling from boys in the group]
>
> Justin: Yeah, explore her and talk to her and learn about each other and you find out about each other, you teach yourself. I can hear people, I dunno man, I think they need a couple of lessons, think they need a couple of lessons.

Justin indicates that 'getting a girlfriend' makes sexual learning possible. Learning from girls, in the context of heterosexual relations, gives young men access to knowledge that is highly prized and based on a 'doing' that enhances masculine identities. Talking to girls and having sex with them consolidates a privileged version of heterosexual masculinity. Here, females are spoken about in terms of landscape; as strange uncharted territory that can be 'explored' and known. Seeking relationships with women in these terms turns sexual learning into a form of territorial conquest which can be incorporated into the masculine repertoire. In this configuration, sexual knowledge is acquired through 'doing' and 'doing' gives young men status which makes them *feel* powerful. This sense of male power and agency in the domain of the sexual has been conceptualized from a feminist perspective as the power to control women. In this context, however, while controlling women may be implicit, Justin uses power/knowledge as a way of controlling other young men. His response to the undercurrent of boys' whispers and laughter implies that experience gained through sexual relationships with women gives young men confidence in the male peer group and the ability to put down others.

CONCLUSION

The consolidation of heterosexual masculinities through sexual activity with young women can be seen in Connell's terms as 'body-reflexive practice'; a circuit where lived experience interacts with societal structures. The recognition accorded to masculine heterosexual activity may imply that there is a comfortable relationship between dominant masculinities and male bodily experience. However, the ethnographic evidence cited in this paper suggests that the links between heterosexuality and masculinity are not 'natural'; they have to be naturalized through practices which incorporate them into a particular version of masculinity. In this version heterosexuality can be viewed as central to an active masculinity premised on doing and displaying. The young men I spoke with indicate that heterosexuality is constantly reinforced within the masculine repertoire through talk and action. For these young men, heterosexuality is *constituted* for them through the practices of sex-talk, masturbation and pornography and *consolidated* in relationships with young women.

REFERENCES

Connell, R. W. (1995). *Masculinities*. London: Polity.
Crosland, M. (Ed.) (1993). *The Passionate Philosopher, a Marquis de Sade Reader*. London: Minerva.
Easthope, A. (1990). *What a Man's Gotta Do, the Masculine Myth in Popular Culture*. Boston: Unwin Hyman.
Foucault, M. (1978). *The History of Sexuality, volume 1*, trans. R. Hurley. Harmondsworth: Penguin.

Guillaumin, C. (1993). The Constructed Body. In: C. B. Burroughs & J. D. Ehrenreich (Eds), *Reading the Social Body.* Iowa: University of Iowa Press.

Haywood, C. (1996). Out of the Curriculum: Sex Talking, Talking Sex. *Curriculum Studies, 4*(2): 229–249.

Kehily, M. J., & Nayak, A. (1997). Lads and Laughter: Humour and the Production of Heterosexual Hierarchies. *Gender and Education, 9*(1), 69–87.

Lees, S. (1993). *Sugar and Spice.* Harmondsworth: Penguin.

Mac an Ghaill, M. (1991). Schooling, Sexuality and Male Power: Towards an Emancipatory Curriculum. *Gender and Education, 31*(3), 291–309.

Thomson, R. (1997). 'It was the Way we were Watching it': Young Men's Accounts of Pornography. Paper presented at the British Sociological Association annual conference, University of York, April.

Thomson, R., & Scott, S. (1991). *Learning about Sex: Young Women and the Social Construction of Sexual Identity.* London: Tufnell Press.

Willmott, P. (1966). *Adolescent Boys of East London.* Harmondsworth: Penguin.

Wood, J. (1984). Groping towards Sexism: Boys' Sex Talk. In: A. McRobbie & M. Nava (Eds), *Gender and Generation.* London: Macmillan.

MASCULINITIES AND ENGLISH PROFESSIONAL FOOTBALL: YOUTH TRAINEESHIP, SUB-CULTURAL EXPECTATION AND GENDER IDENTITY

Andrew Parker

INTRODUCTION

Viewed either in terms of its occupational or social characteristics, professional football is a strictly gendered affair. Its relational dynamics, its working practices, its commercial ventures, its promotional interests, are replete with images of maleness (cf. Williams, 1994; Williams & Taylor, 1994; Williams & Woodhouse, 1991). Focusing on the process and practices of Youth Training within the confines of one English professional football club – Colby Town – this chapter draws on ethnographic research into youth training within the professional game and looks at how the everyday process and practices of occupational indenture served to shape the masculine identities of the trainees concerned in accordance with both official and unofficial institutional norms, standards and expectations.[1] The theoretical concept of 'hegemonic masculinity' (cf. Connell, 1987) is used to analyze and categorize the construction of a range of institutional masculinities. Additionally, the notion of working-class, shop-floor culture forms the back-drop against which trainee relations and

Genders and Sexualities in Educational Ethnography, Volume 3, pages 41–65.
ISBN: 0-7623-0738-2

actions might also be viewed and located (cf. Clarke, 1979; Collinson, 1992; Willis, 1977, 1979). What transpires is a broadly descriptive analysis within which trainees are categorized according to their divergent lifestyle behaviours, their commitment to occupational success, and the specific masculine traits which they exhibit.

CONTEXT AND METHOD

Colby Town is a prosperous English professional Football League club which, over recent years, has built up a formidable reputation for its nurturing and development of young players. During the 1993/94 season it supported a youth team squad of 20 players and a professional playing staff of approximately 35. Of the young players at the club, eight were first-year Youth Trainees, 11 were second-year Youth Trainees, and one individual, although officially recognised as a full-time professional player, was eligible for youth team selection on account of his birth-date. All first-year trainees were between the ages of 16 and 17 and had arrived at Colby straight from school. Accordingly, second years were embarking on their second full year of paid work after leaving compulsory education, and were all between the ages of 17 and 19.

Trainees were interviewed at least twice over the course of the research period which lasted the full duration of the 1993/4 season – from early July 1993 until May 1994. For the most part I attended the club for three days each week as participant observer, spending two days training, working and socializing with trainees, and one day at local colleges of Further Education as a fellow student. After the initial three months of participant observation I began to conduct interviews with trainees in the privacy of Colby Town's residential youth hostel. The vast majority of these interviews were carried out on a one-to-one basis. On only one occasion, towards the end of the fieldwork period, did a group interview take place in accordance with the collective requests of seven second-year trainees. Interviews were also conducted with various members of club staff (i.e. youth team coach, Education Officer, youth hostel proprietors) and the college tutors directly involved with Colby trainees on day-release courses. All interviews were tape-recorded. To supplement this data, a detailed fieldwork diary was kept throughout the research period. In an attempt to limit institutional suspicion, this was written-up each evening on return from club, college and/or social settings. For a more comprehensive methodological account of youth traineeship at Colby Town see Parker (1996a, 1998).

MASCULINE CONSTRUCTION: INSTITUTION AND INDIVIDUAL

Whilst English professional football is primarily a male preserve, it would be both inappropriate and inaccurate to discuss the game's heavily gendered complexion in relation to the existence of a singular, monolithic masculine form. Instead, professional football clubs (like a host of other sub-cultural settings) should be viewed as institutions which exhibit a dominant or hegemonic masculine form, beneath which a hierarchy of alternative subordinate masculinities continually challenge and contest this pre-eminent position (see Connell, 1987; Mac an Ghaill, 1994).[2] In a general sense hegemonic masculinity, Connell (1987) argues, should be seen to be constructed in relation to (and in opposition to) femininity and subordinated forms of masculinity and is largely characterized by notions of heterosexuality, power, authority and competitive aggression. It is worth noting, however, that each institutional locale is likely to exhibit its own specific brand of hegemonic masculinity which, whilst incorporating these general features, may well emphasize particular contextual idiosyncrasies and nuances.

Given this 'theoretical' pre-disposition, how might we explain the formation of trainee gender identities at Colby Town? The hegemonic masculine ideals in place regarding the actual club setting were defined in terms of an explicit institutional logic which incorporated notions of personal integrity, conscientiousness, discipline and the development of a healthy 'professional attitude'. Such values, in themselves, strongly reflect a masculine working-class legacy which has come to shape the historical contours of English professional football (cf. Clarke, 1973; Critcher, 1972; Taylor, 1971a, b). In particular, the whole notion of 'professional attitude' held specific importance in terms of how well trainees were seen to accept traditional working practice in that its assessment was based around the extent to which individuals accommodated both the routines of occupational duty (i.e. the fulfilment of menial/domestic chores around the club) and the physical and psychological rigours of actual performance both in training and youth team fixtures. Menial/domestic chores included the neat laying out of professional kit, the cleaning of professional and trainee boots and the cleaning and servicing of training ground equipment. Additionally, first-year trainees were compelled to carry out the more general daily cleaning tasks within the confines of the club.

It was crucial that all trainees demonstrated a keen and 'hardy' enthusiasm for the game itself, a forceful 'will-to-win', an acceptance of work-place relations based on authoritarianism/subservience, an ability to conform to institutional ('official') values and disciplinary codes, and a commitment to social and professional notions of solidarity and cohesion. Providing trainees readily adhered to these stipulated norms and values, and providing also that they developed and matured to expected levels of footballing competence, they then stood a reasonable chance, it seemed, of completing successfully the transitional phase from Youth Trainee to professional player.

Although levels of enthusiasm and motivation fluctuated amongst youth team members as regards their overall attitude towards the rigours of training and playing, central to the occupational identities of all was a general commitment to a successful career in football and a psychological acceptance of institutionally defined hegemonic masculine requirements. In conjunction with these ideals, however, traineeship was also lived out in relation to a number of 'unofficial' behavioural norms and values which collectively impinged upon individual masculine constructions. Issues and experiences concerning heterosexual relations, wealth and consumption, for example, all had a part to play in this respect (see also Griffin, 1993; Hollands, 1990; Willis, 1990). Added to this, the hyper-masculine behaviours of a selection of Colby's professional players also had a significant (if distanced) impact on trainee life patterns. What this meant in terms of the formation of individual masculine identities was that whilst being obliged to consider the occupational standards espoused by club officials, trainees were also under obligation to adhere to the social expectations of the all-male subculture within which they lived and worked. Because degrees of individual investment within 'official' and 'unofficial' behaviours varied, these somewhat contradictory circumstances necessarily led to the generation of a range of divergent trainee masculinities, a closer examination of which I now undertake.

CONSPICUOUS CONSUMPTION: CLOTHES, CARS AND 'CASH TO SPARE'?

As a consequence of their continual emulation of professional player lifestyles, a key leisure-time pursuit for trainees at Colby Town was to enter the consumer market-place. Here three main areas of consumption dominated individual spending: clothes, cars and socializing. Together these interests formed a series of tightly interwoven linkages around which aspects of individual autonomy and identity were established via a matrix of assumptions concerning the establishment and fulfilment of masculine prowess, kudos and reputation.

According to popular peer group belief, for example, the ultimate trainee social pursuit of 'women' and (heterosexual) 'sex' could only be carried out to its full potential if and when one could boast: mobility via car ownership, an extensive range of designer clothing, a healthy cash-flow position, and frequent appearances 'out on the town'. Cumulatively, these lifestyle features necessarily allowed a demonstration of wealth, style and consumptive power which readily enhanced individual images of maleness.

Considered within the broader context of youth culture such trends are not atypical. Hollands (1990, p. 150), for instance, has highlighted the way in which, despite their lack of disposable income, "clothing was the major item of consumption" for the Youth Trainees within his study. Similarly, in reporting the findings of their survey of young people in Scotland, Hendry et al. (1993, p. 126) have stated that not only are "appearance and fashion" likely to represent common elements of peer group status amongst adolescent males but, in addition, such lifestyle components may also be related to class origin in terms of style replication (see also Clarke, 1976; Hall & Jefferson, 1976).

As far as trainees at Colby were concerned, contemporary dress-sense was tantamount to the development of an acceptable masculine identity. Clothes spoke implicitly of the individual beneath them and were looked upon as an indication of the extent to which individuals adhered to informal group norms. Insofar as professional footballers (within England at least) have long since regarded themselves as influential trend-setting agents (cf. Chapman, 1993; Gowling, 1974; Taylor & Ward, 1995), the purchase of exclusive designer clothing for boys at Colby symbolized not only their levels of wealth and peer-group standing, but their desire to be associated with the game's élite male order. First-year trainees in particular felt the pressure of fashion conformity more than most.

> AP: There seems to be an identity thing going on around clothes; do you feel under pressure to buy certain kinds of clothing?
>
> Charlie: There probably is. I think the lads buy clothes that are very similar, but they buy them, and they're like, 80 or 90 quid for some. You tend to, like, buy pairs of jeans that you can buy from 'Next' or something for about 35 quid, but they go and buy some for 90 [quid] or whatever.
>
> AP: What sort of 'makes' are you talking about?
>
> Charlie: 'Armani' [Giorgio Armani] an' stuff like that, 'Ralph Lauren'. I've got 'Armani' ones on now. Look. It's just silly money really and it's just daft, and you just do it 'cos . . . I mean you look good. You look good in the same stuff . . . without the 'Armani' stuff on, but its just because its got the 'Armani' [motif] on that I feel a bit more confident. I know it sounds daft but . . . . Its the same when you've got a [designer] shirt on or whatever, it just makes you feel a bit 'bigger'. That's the way it is. It's just daft money. I wish I could have the self-control to try not to spend that sort of money.

Crucial to first years was a need to match the precedent set by older trainee group members.

Charlie: I think the second years are very much image conscious . . . . The clothes thing, it's the second years that's got the clothes, 'cos if the second years didn't have 'em, we wouldn't get 'em. But I think the clothes is just a way of showing people that we've got the money. Its just a daft image thing I reckon.

Adrian: The second years . . . they've all got nice clothes and everything, and I think that when we [first years] first came here, probably still now, they've got a lot bigger wardrobes than us. They'll 'rip' us sometimes for like, "Oh no, I can't believe you've got those jeans on," and really, y'know, that's intimidating 'cos, y'know, we'd like to have clothes like that, y'know, expensive clothes. But . . . it's not easy when you've been on YT money and then you're, y'know, trying to – I suppose you're copying them . . . clothes are just so expensive what we buy, and I suppose sometimes you just feel under pressure that you've got to wear, [or] buy the same clothes – 'Ralph Lauren', 'Replay' jeans, 'Paul Smith', 'Stone-Island', things like that, and 'Hugo Boss' stuff. It's ridiculous really. I mean, I think these 'Paul Smith' jeans were 50 quid or 60 quid. In a way it's like when we were at school, you try and look you're best – well, we tried to look pretty smart, and if one of the lads got a nice pair of shoes, we'd all copy. It's just the way life is really and it's to do with who you associate with.

Nor did informal controls over standards of dress relent in the case of everyday casual and work-place clothing. If daily apparel did not come from the sporting repertoires of specific companies (i.e. Nike, Reebok, Adidas, Umbro, Mizuno, Puma), then it failed to gain peer group approval. In this sense, 'sub-cultural style' held considerable sway, to the point where consumer culture, 'the look' and the fashion-based expectations of various sporting industries clearly dominated trainee appearance (see also Featherstone, 1991; Hargreaves, 1987; Parker, 1996b). For more detailed accounts of the relationship between football, fashion and youth culture see Redhead (1991) and Giulianotti (1993).

Despite recognizing their own naiveté in consumerism, many boys found it hard to curb spending habits. The wage structure of the club itself did not help in this respect. To escape YT managing agent restriction concerning recommended levels of trainee pay, first years were allowed to remain on official YT contracts for the first 12 months of their employment only.[3] Once 17, they became eligible for professional player status and were thus transferred to professional contracts for a further one-year period to legitimate increased financial income. What this meant in terms of actual earnings was that prior to their 17th birthday (and in line with national recommendations) first years received £31.50 per week with no charge for food or lodgings at the club's residential hostel. Trainees aged 17 received £160 per week with £225 per month being deducted to cover accommodation fees. Those aged 18 and over received £200 per week with corresponding monthly deductions. In addition, all trainees received some kind of signing-on fee from the club which was paid in three equal instalments during the course of the trainee period.

These fees were negotiated on an individual basis and ranged from £1,600 to £4,000 per trainee.

Hence, coming-of-age financially brought with it a host of pressures and obligations which had to be fulfilled if peer group acceptance was to be maintained. Besides the purchase of clothes and cars, turning 17 marked the point at which trainees were expected to associate and socialize more readily with reserve (and sometimes first) team players – to copy their behavioural trends, to frequent their social venues, and to emulate their general lifestyles. No longer was it enough simply to enact the social status of first year 'skivvy'. From this point on peer-group belief held that trainees should take it upon themselves to bolster their own identity via a pursuit of more 'professional' pastimes (i.e. regular drinking sessions with team-mates, [hetero]sexual exploits and promiscuity, daily visits to 'the bookies') all of which contradicted the clean-living intentions of the club (cf. Wilshaw, 1970). Speaking during an end-of-season interview about the social experiences of traineeship in general, second-year Gary Riley described how youth team drinking habits in particular appeared to reflect a kind of masculine 'occupational inheritance' of professional player behaviour.

> Gary: I'll be honest with you right, it is true . . . . It's just the lifestyle. Because what you're doing, you've not got the chance to go out like people who've got normal jobs. They can go out and have a couple of pints at dinner, and a couple after work. Here, once you do decide to go out on the piss, you go out on the piss and you get pissed, and you drink like fuck. Because it's so highly pressurised. Sometimes you get so depressed you just want to go out and get steamin', and you do. I mean I've been steamin' a few times up here. I don't know why it is but everybody, well, the majority of footballers, do drink heavily, I would say with a few exceptions. Especially at this club, all the young lads drink like fuck.

Smoking too, was all part of an image construction process into which the majority of trainees were keen to invest at some time or another.[4]

> AP: When they [trainees] go out does everybody like a smoke as well?
>
> Gary: Well a lot of them do. At weekend . . . there's about two people out of the whole 19 of us that don't.
>
> AP: And is that for the same reason, that you can't do it, that it's restricted here?
>
> Gary: I don't think so, its just 'cos when they're pissed they do it. I mean I do it anyhow. I do it all the time 'cos I used to smoke when I was at home. Like, I do when I go out for a drink up here, but at the weekend I do it properly again all the time . . . .
>
> AP: But how much of that is about identity as a footballer . . . . Is there something 'good' about smoking a lot or drinking a lot?
>
> Gary: Yeah, I think it is definitely. 'Cos when we first came up here . . . everyone was like, "Oh you're a disgrace, get out of my room, don't talk to me," and stuff like that, as though

it was something out of the ordinary. I mean now if I go out with a pack, everyone's "Oh give us one of them" – I think its just the image really . . . when they're having a drink I think they just like to do it for effect. They're not enjoying it because they get up in the morning and they say, "Fuckin' hell, my hands stink, my breath, I can taste it, it's fuckin' horrible," and things like that. But it is definitely, yeah.

Given the competitive pressures surrounding trainee life and the enthusiastic approach which individuals employed towards social endeavour (cf. Gowling, 1974), it was perhaps not surprising that alongside feelings of overall merriment other less positive consequences of nights-out emerged. Common, for example, were bouts of vomiting in the early hours of the morning, hangovers during training, and personal vows never to do such things again. Worse still, for some this social catalogue of smoke, drink and spending machismo regularly transcended the realms of fun and enjoyment and turned instead into an excuse for displays of aggression and violence which at times implicated the rest of the trainee group.

Sentiment towards such occurrences differed greatly. Whilst some regarded 'social' violence as a form of amusement, others put it down to gross immaturity. Either way, far from constituting a general trend, behaviour of this nature was predictably attributable to particular parties – namely second years Robin Hindle and Ben Tattersall – both of whom were renowned amongst the youth team group for their lack of self-control in social situations.

Colin: Like Robin, I mean he's so funny when he goes out. I mean, he's brilliant. Y'know he'll have such a laugh, and he'll 'take the mickey' out of people. But like some people take it the wrong way so he'll end up getting in a fight or getting into trouble. But that's the way he is. He doesn't do it on purpose, he just does it like having a bit of a laugh.

Jimmy: I think . . . they want to be noticed, sort of thing. Probably 'cos when they've been at their old school they've probably been good at everything. Y'know what I mean, they've been like 'top-man' an' that, an' they've come here an' everyone's like that. Y'know what I mean, an' I think they can't handle it. They want to be the best . . . they want to be when everyone says "Oh yeah, he's a good laugh," an' stuff like that. They just want to be noticed don't they . . . an' all they want to do when they go out is get a few drinks down them and they just want to fight. Like a couple of times we've been out, all the lads, they have three or four pints and they just want to fight. Bloody stupid.

Described by first year Charlie Spencer as the team 'nutcase' who always came back 'smacked to bits at the weekends', Robin Hindle demonstrated his aggressive tendencies early on in the 1993/94 season by head-butting first year Andy Higgins during an argumentative night-out in town. Leading to serious repercussions for all trainees, this incident not only threw Higgins's stay at Colby into question on account of his own personal discomfort, but it resulted also in an official club warning for Hindle and a blanket reduction in Saturday night curfew times from 12 midnight to 10.30 pm.

With regard to general patterns of trainee consumption and behaviour, it is important to point out that although drinking trends did appear to follow professional player influence, smoking and violence in social spheres equated less obviously with instances of what might be termed 'occupational inheritance'. True, trainees had witnessed the outbreak of drunken scuffles amongst professionals at internal club functions (i.e. Christmas and end-of-season parties), as they had the occasional smoking of cigars and cigarettes. But by and large these latter habits seemed to mirror the traditional expectations of working-class adolescent life and contemporary youth culture far more than the cloning of professional player conduct (see, for example, Canaan, 1996; Hollands, 1990; Parker, H., 1974; Willis, 1977; Willmott, 1966).

Yet such habits were not the only ones to feature prominently within the social lives of Colby trainees. Crucial also to individual and group outings, and to the overall establishment of masculine prowess, was the pursuit and exploitation of heterosexual relations – a topic never far from peer-group discussion.

HETEROSEXUAL RELATIONS: GIRLFRIENDS, 'GROUPIES', 'BABES' AND 'RIPPERS'

Girls were something of a contentious issue for Colby Town trainees. Getting a girlfriend was, for the majority, relatively unproblematic. Working out whether or not the club actually approved of such relationships, was another matter altogether.

> AP: Do they ever say anything about girlfriends or anything like that?
>
> Steve: No. I mean, it would be better if they told us whether like we could [have them] . . . I meant lads they don't know . . . I mean we don't see any girls like hardly anyway, but if we did we don't know whether we'd be allowed one or not, I suppose they can't stop you havin' one, but . . .
>
> AP: What, you mean in Colby?
>
> Steve: Yeah, I mean they don't tell you. They don't sort of encourage you, but they don't discourage you, y'know what I mean. They should let you know one way or t'other I think.

Central to this state of confusion were two contrasting beliefs concerning heterosexual relations to which all trainees were exposed either prior to, or during, their time at Colby. On the one hand, girls were located by staff/officials as "the root of all evil," and as facilitators of occupational failure due to their distractive potential in terms of mental and physical commitment (see also Hopcraft, 1971; Sabo & Panepinto, 1990; Shankly, 1977).[5] On the other hand, 'steady' girlfriends were regarded by some as beneficial to the lives of young players in that they allegedly provided a calming social influence which necessarily reinforced notions of discipline, loyalty and personal responsibility (see Best, 1990; Brooking, 1981; Douglas, 1973; Gowling, 1974).

Both sides of this moral dichotomy were evident amidst trainee opinion. Revealing the existence of this latter viewpoint, for instance, two boys involved in long-term relationships, Gary Riley and Davey Duke, mapped out their understanding of the rationale behind it.

> Gary: I mean I think it does you the world of good having a steady girlfriend, 'cos it stops you messing around and wanting to go out all the time, and you've got something to look forward to at the weekend. That's how I see what I'm doing, 'cos I look forward to seeing my girlfriend at the weekend. Whereas if you haven't got that, you just want to go out and get loads of different girls, don't you. And that's when you get yourself in trouble messing around, when you start nicking other lads' girlfriends – that's when you start getting your head kicked in.

> Davey: It does keep you out of trouble, y'know what I mean. It stops you goin' out with your mates an' stuff like that. But they [the club] probably don't see it that way. They just probably think you should just concentrate on your football. But I think that the ones with the girlfriends up here are the more sensible, y'know what I mean, the ones that don't go out drinkin' every night.

Because the majority of trainees professed to having girlfriends, either in and around Colby or within the vicinity of their home-towns, such inferences formed the accepted basis of general peer group belief. Of specific concern to many, however, was the way in which a more critical view of girls surfaced in and through the informal interactive messages of the work-place. A central figure in this respect was youth team coach Terry Jackson.

> Gary: He [Terry Jackson] always used to say, "Oh, you're going home to see Jen are you?" – all the time like this, and I just used to say, "Yeah, I am." And he always used to give me stick about it but he just doesn't bother anymore. He's just give up now 'cos he knows he can't do anything about it . . . . Jimmy, who's like one of his 'mates', . . . he's just started going out with a girl back in Scotland, and Terry gives him a bit of stick about going home and seeing her every weekend, and how he should be stopping here 'reviewing his game' instead of boning his bird all the time . . . . You can tell he doesn't like you having a girlfriend. He wants you 100 percent committed. Nothing else to stand in your way.

> Nick: It's int' corridor, it's never a sit down talk . . . about girlfriends. He'll [Terry] ask, "Did you give your lass one this weekend?" an' things like that. An' 'Oh you want to kick her out o't winda if she don't want nowt t'do wi football', an' things like that. That's what he's like wi ya about relationships. I don't think he minds us seein' girls or owt like that, its just that he 'teks the mick' . . .

Local boy Martin Walsh found the personal nature of such inference particularly disconcerting.

Martin: He [Terry] seems to think that if you've got a girlfriend, you're married. But my view on it is if you've got a day off or an afternoon free or something like that, you take your bird out or you go to the cinema or something like that. It's not as if you're fuckin' married, "Oh yeah, I'm comin' to your house every bastard day, I'm gonna make you breakfast in the morning," and things like that. He thinks you're movin' in together. He don't like it, y'know, 'cos he thinks you're just gonna think about your bird all the time.

AP: Has he said that to you?

Martin: Yeah. That's what I mean, he gets involved in your personal life and that's nothin' to do with him. If I want a girlfriend or if I don't it's nothin' to do with him.

AP: Well, what does he say to you?

Martin: He just said, "Oh you've got to show her who's boss," an' all this and, "You're under the thumb," an' all this shit . . .

To some extent these data might serve to reflect little more than a genuine professional desire on the part of Terry Jackson to monitor closely individual levels of occupational dedication, enthusiasm and commitment, whilst outlining one of the ways in which he sought to make himself more accessible in terms of coach/trainee relations. What they fail to articulate, however, is the defining limits of club tolerance as regards girlfriends in general, an issue which I took up with Jackson during interview.

AP: What about stuff like girlfriends. Does the club have a policy? Does it . . . mind?

Terry: Well sometimes. It's down to the individual. Sometimes we've got lads that have got very steady girlfriends. We've got young lads that haven't got any girlfriends. They're all different. As long as they know where their priorities should be. I mean at the end of the day their priority should be trying to make a profession in the game. I always think if they do everything they can with their football, and they work very hard during the week, then let them go out at the weekend, let them enjoy themselves – go out with their girlfriends and have an enjoyable weekend and get a little bit of relaxation which young lads need.

AP: There seem to be two schools of thought about if they have a girlfriend it might distract them or it might quieten them down?

Terry: Well, I think what suits one doesn't suit the other. We've got totally opposite lads here where some . . . have got girlfriends since they've been at school, and some lads that don't want that, so again its an entirely individual choice.

AP: So there's no policy on it then?

Terry: No. No policy at all. The only policy we have is if it's getting a little bit too heavy and they're not concentrating on their football and it's all the other side, then we'll crack down. But otherwise we just leave it alone to the lads. I mean at the end of the day they are young men.

Though structured around a liberal rhetoric of 'subjective choice' there are, of course, shades of official prescription here as regards the boundaries of trainee relational activity. Added to this, and implicit within Jackson's more general everyday assertions, was the distinctive location of women as a potentially contaminating force against which trainee careers must be protected. Arguably, such codes were intimately linked to broader underlying assumptions concerning the way in which young footballers may well be viewed by girls as a 'good catch' on account of their affluent and prestigious occupational position and/or potential. This scenario, it seemed, left Colby trainees in continual danger of being 'trapped' within a highly charged emotional nexus of 'loose' female sexual advance and 'natural' male desire.

Amidst these pervasive, if sometimes quite vague, elements of institutional belief, were reflections of what Hollway (1984, p. 232) has described as a 'male sexual drive discourse'. Portrayed as 'beings' dominated by an innate reproductive urge, within such postulations men are defined as sexually incontinent and out of control and thereby 'naturally' excused their frequency and method of sexual approach. Women, on the other hand, are depicted as sexual subordinates and/or objects of male desire who in embracing this 'object' position are thus located, Hollway (1984, p. 232–3) suggests, as "men 'trappers' via their powers of sexual attraction."

The existence of such discursive practices at Colby meant that player popularity and media attention were implicitly framed as hazardous aspects of 'professional' life. Public exposure was seen to create the conditions under which women might represent some kind of sexual 'attraction' to trainees with players themselves possessing their own 'attractive' qualities in the form of occupational prestige, financial gain and social standing. Hence, whilst fostering a desire to entice female company in accordance with the all-male peer group relations in play, club culture also engendered an inherent fear of heterosexual intimacy on the grounds of exploitative risk.

So intense were Terry Jackson's feelings that in addition to informal verbal manifestation his own protective measures were occasionally displayed through explicit gesture and action.

> Martin: He's [Terry] just like a kid . . . . He can't relate to people . . . . If we went on tour, or we went to another place, or at the ground, like if a bird walked in you'd go "Woh! Woh! Phore!", wouldn't you, Fuckin' hell, what a bird, an' all this, an' Terry would think, 'Oh fuck', he'd go red and he couldn't talk to them . . . . Like the insurance bird [female insurance representative] comes in, an' she's a bit of a 'sort' like, an' she's walking up an' she says, "Oh, can I see the lads about their insurance?" And he can't look her in the face and say, "Look, this is no, like place, 'cos the changing rooms are there an' everyone's walking round with their nobs out an' that – do you mind just waiting outside." He has to say, "Look, just get out!" He can't talk to her . . . and by doing what he does to the lads, like keeping them locked-up an' that, he doesn't get any respect. And that's why when he says things to them, they just think, 'Prick'.

AP: But maybe he thinks that if he did trust you, you'd let him down?

Martin: But he doesn't give us a chance, which he should do, 'cos there's only one way of finding out, isn't there.

Insofar as popular representations of professional players occasionally attempt to sensationalize 'an association with women' as one possible aspect of individual downfall (see, for example, Best, 1990), there may be some grounds upon which the attitudes displayed by Terry Jackson could be seen as a worthwhile consciousness-raising exercise. Trainees were, however, already aware of such relational pitfalls. Martin Walsh, for example, went on to explain how, as a result of his own negative experiences, he had come to scrutinise more closely male/female relations particularly in terms of personal exploitation.

Martin: I don't just go for a bird just 'cos she's gonna sit on my cock, y'know what I mean . . . I'm not gonna go with some bimbo like some of the lads. I mean it's just the talk i'nit. A lot of it's bull-shit what they come out with . . . .

AP: Do you think there is something about this footballer and . . . [interrupted]

Martin: Footballer – birds, yeah. It goes don'it – I mean you get labelled 'footballer' and just 'cos you've got money in your pocket, probably a flash car, and then some of the footballers who do let what they've got go to their heads, they do take it to their advantage . . . .

AP: Do you find it then at your age?

Martin: At my age yeah. You get like stupid little girls an' that, but I mean the lads love it, all the slags an' that, they love it. I mean I have a laugh, I mean yeah there is birds but you've just got to be careful ain't ya. I mean 'cos you don't want nobody coming back at you, do you, to be fair. I mean, if I go somewhere like Manchester with mi mates, I'm someone else me, same as I am on this fuckin' tape. I'm Billy Bunter for all they fuckin' know. I mean they don't give a shit do they. But round here I've got to be careful fuckin', 'cos its mi home town, an' if mi bird found out its end-of-story i'n'it. But I mean I like mi bird, an' there's plenty of other birds . . . but I don't use mi bird, I don't just shag her like all the other lads who just want a bird for a shag, 'cos I get on with her.

AP: And you know the same with her?

Martin: Yeah, that she's not going out with me just because I play football.

As well as demonstrating the existence of some kind of protective emotional device, Martin Walsh provides clear evidence here of the more general way in which trainees at Colby both viewed and verbalized the details of their associations with girls. In accordance with other examples of male-centred working-class adolescent life, a distinct form of trainee chauvinism existed whereby girls were predominantly portrayed as being utilized purely for the sexual fulfilment which they might provide (cf. Jenkins, 1983; Parker, H.,

1974; Walker, 1988; Willis, 1977; Wood, 1984). These attitudes were accompanied by forms of male-centred jargon which sought to describe accurately the sexual and/or physical attributes of females. Within this system of trainee categorization girls and young women were generally referred to as 'birds', 'tarts' or 'chicks' (all of which served to demean and objectify women in some way) irrespective of the depth of their association with youth team members. The only real exception in this sense was the prefix 'classy', which informally signalled some kind of professional or socially mobile demeanour on the part of the females concerned. In turn, girls making explicit advertisements towards their sexual availability were commonly known as 'slags' whilst those wishing to remain distant from intimate sexual involvement were rendered 'tight fuckers'. Likewise, just as girls who possessed attractive facial and/or bodily features were termed 'babes', those failing to match such group-imposed standards were accorded the more derogatory label of 'dogs' or 'rippers' (see also Cowie & Lees, 1981).

Further connotations can also be drawn from the words of Martin Walsh as regards broader youth team attitudes towards heterosexual relations. Whilst the majority of trainees at Colby were in the position of either having a girlfriend or not, some boys adopted an intricately balanced double-standard, accommodating a combination of 'steady girlfriend' and sexual promiscuity (see also Griffin, 1993; Jenkins, 1983). Player possession of 'a bit on the side' or 'a bit of spare', it seemed, went a long way towards the individual fulfilment of institutional expectations surrounding the creation of an acceptable masculine identity.

> Neil: . . . most of the lads e'll say, y'know, you say you're a footballer and you'll get the 'chick' and use her and that's it. I mean, when I was talking to those lasses in college that lass that I fancy, she goes, "I know you's are all footballers." I said, "What do you mean by that?" and she wouldn't tell me. Well what she meant was that you just go with one and then go with another, and you're all over the place. All the women say about footballers is that they just want women and then finish with them . . . .
>
> AP: But is that not something other young lads do?
>
> Neil: Well I mean they'd like to wouldn't they, but because we're like footballers and we'll pull the 'chicks' an' that we'll probably get called more of a slag. Like if you go with more you get called a slag. But if you go with one or two you won't get called that much. But footballers seem to go for loads don't they, 'cos of who they are.

Interesting here is Neil's use of the term 'slag' in relation to trainees themselves. He was of the unique opinion that to be known as a male 'slag' was in fact a relatively undesirable position to be in, particularly with regard to one's future potential as a prospective boyfriend. Conversely, most trainees believed that the worse their reputations with girls became, the more likely they were to attract additional female company. The possible achievement of these broader aims was further reinforced by the fact that daily club training sessions were often observed not only by an array of curious adult spectators but also by cohorts of school-aged girls known amongst youth team members as 'groupies'. Alongside myths of footballing virility, this highly visible female presence represented an example of the way in which trainees were a much sought-after commodity within the teenage heterosexual market-place. In fact, so enthusiastic were a number of these girls, that as well as regularly attending youth team fixtures they devoutly frequented the club each weekday lunch-time in order to catch a glimpse of their trainee heroes, and/or to pass on affectionate notes, cards and small gifts to them.

Ironically, because intimate physical encounters with 'groupies' were only considered a possible option if and when trainees "couldn't get a shag anywhere else," on the whole these recurrent habits merely served to reduce squad feeling towards their reverent fans. Nevertheless, as far as wider social conquest was concerned, such occurrences did provide some measure of the extent to which trainees' status could be utilized to command female attention – an occupational benefit which individuals were only too happy to sample in terms of their own levels of social exploitation.

> Nick: Like its good because you're a footballer and you've got that name of footballer which you can flash about . . . it's being macho, y'know what I mean, that's what it is. 'Oh, I'm a footballer' . . . and you can use it to come across to people an' that.
>
> Jimmy: They go out and say, 'I'm at Colby', an' all that. When I go out it never comes into my mind to tell everyone that. It might do, if they ask you what you do I don't lie – sometimes I have, 'cos it's not the right time or the place – but other times you say, "I play football." Y'know what I mean, well it's not worth lying is it, especially if it's a nice bird, and there's a chance of a 'pull' mate, tell her [laughter].
>
> Colin: People put you in the same place as though you're a superstar an' its good really. I mean it might sound as though I'm being big-headed but its true . . . and you'll get like you're girls an' whatever, an' they'll think you're some sort of superstar. Its great, an' you have a bit of a laugh, an' you go along with it, don't you. But its good for your social life I must admit.

Whilst interview discussion with trainees raised a host of issues surrounding both the negative and positive aspects of heterosexual relations, it also facilitated an indication of the way in which sexual activity was framed within the context of wider work-place relations. In this sense, the manifestation of chauvinistic codes and attitudes within the personal lives of trainees symbolized not only the inferior social position accorded girls and girlfriends, but also the ingrained depth of male sexist practice at the club. In constituting a key aspect of identity formation amongst trainee and professional players alike, females (and their associated worth) featured as a fundamental element of everyday life in this respect. Moreover, the location of women as subordinate to men within the hierarchical bounds of hegemonic gender relations was central to the conversation of the work-place, particularly in terms of the shop-floor cultural discourses employed.

MASCULINITY, HUMOUR AND 'WORKING-CLASS SHOP-FLOOR CULTURE'

The constituent elements of the occupational culture surrounding working-class shop-floor life have been well documented (Clarke, 1979; Cockburn, 1983; Collinson, 1988, 1992; Tolson, 1977). Within this literature manual labour has been cited as a collective whirlpool of informal work-place relations predominantly comprising a strict male chauvinism, a 'breadwinner'/manual production mentality, and a coarse sexist humour manufactured around practical jokes, gestures and racist/homophobic connotation (cf. Beynon, 1975; Collinson, 1988; Gray, 1987).

In accordance with these inclusive elements, the foundations of occupational culture at Colby were grounded predominantly within the realms of sexuality (and to a lesser degree ethnicity), and were made manifest in and through the intricacies of institutional language and interaction. Central to the enactment of this cultural lifestyle was the stylized adoption of a sexually explicit and often highly derogatory vocabulary which was ideally characterized by a sharp-pointed form of delivery.

Put into practice within the context of relational work-place humour such language took the shape of the previously well documented process of 'piss-taking' – or 'ripping' as it was more commonly known amongst Colby Town players (cf. Chapman, 1993; Clarke, 1979; Collinson, 1988, 1992; Lyman, 1987; Riemer, 1979; Willis, 1977, 1979). Here, in order to gain any kind of peer-group credibility individuals were not only required to 'take' the insults of others, but to 'give' as good as they got, thereby proving their masculine worth (see also White, 1971). To this end, 'taking the piss' was all about administering verbal 'wind-ups' to the point where work-mates failed to cope with the pressures in hand and ultimately snapped. In separate interviews carried out towards the end of the 1993/94 season, I asked a number of trainees how they had adapted to the details of such informal methods of working practice.

> Neil: You've got to have thick skin haven't you. It just goes in one ear and out the other one most of the time.
>
> AP: But did it get to you at first?

Neil: At first it did, yeah. But y'know, they take the piss out of you, you take the piss out of them, that's the only way to do it really. I really don't mind. I mean, some can't take it.

Adrian: Even sometimes now I get fed up with some of the lads. But y'know, you've got to 'give it' and you've got to 'take it' as well. Like when people started calling me 'bush-head' with my hair to begin with – 'cos it was neither short or long – I was dying for it to grow. I still get called it now, but now it's just a nickname, whereas to begin with it was personal. It doesn't bother me now, but I struggled with it round at the ground a little . . . . There are occasions when some people are in worse moods than others, and when somebody's in a bad mood everybody will wind them up and they'll get mad, but you've got to give and take it. But I think some people take it differently. I mean, sometimes, y'know, when it's not your day, people wind you up, but it's all part of it isn't it. Everybody gives it out; you've got to try your best to take it.

Later;

Gary: Everyone just gets 'ripped' [ridiculed] to bits off one another. You've got to get on with that and not take things to heart really.

AP: Do you think that's all part of football?

Gary: Definitely. Everywhere you go, it's got to be the same. It's just a 'ripping society'. Everything – you get the rip took out of you for everything.

AP: How do you cope with that personally?

Gary: I didn't used to cope with it very well at all really at first, but I'm not so bothered now. I used to get dead mad and start lamping [hitting] people all over the place . . . . Somebody would just say something daft which you would laugh about now, and I just used to get really daft and say, "Let's have a fight then." But now you just get used to it and laugh it off, and call somebody else a name back.

As well as being centred around the details of verbal comment, practical jokes were also a regular feature of trainee life in both occupational and social settings. Whilst within the confines of the club's trainee hostel the ransacking of beds and the hiding of personal possessions were favoured habits, common in the work-place was the random dousing of trainee underwear with 'Ralgex' and the filling of shoes with talcum powder. At the same time, urinating in temporarily abandoned cups of morning tea was not unheard of nor were changing-room mock-battles using tea, sugar, talcum powder, and/or antiseptic cream as ammunition.

Sexual connotation, women, sexual fantasy and derogatory sexist comment were elements around which trainee 'piss-takes', 'wind-ups' and practical jokes were often constructed. To this end, highly spurious stories existed regarding a host of female figures visibly recognizable to the trainee group as a whole. Everyone 'knew', for example, that the 'bird' who came into the club to sell insurance had 'fucked all the first team' during the previous season. Likewise, the alleged promiscuous exploits of the female staff in the club restaurant and laundry were 'common knowledge'. On a more personal level sexual fantasy often underpinned 'humorous' conjecture towards familial relations. Particularly popular in this respect were verbal attacks on girlfriends, mothers and/or other female relatives.

> Gary: We have big slanging matches . . . where people's mothers come into them. As you well know, mothers and girlfriends are a popular target . . . . I mean, some of the things people come out with up here are outrageous. I mean, if you were back at school and someone said it you'd smack 'em. I mean, you'd be having big fights about it an' everything, 'cos people just come up to you an' say, "Oh your mum's a fat slag, mate," an' you're like, that [gestures state of shock]. An' if someone had said it to you for no reason at all in a different job or at school or something there'd be big brawls an' everything. But I think that's just a way of life really in football.

Of course, contrary to this latter inference, Colby Town should not be seen as an isolated institution in terms of its facilitation of such a matriarchal sexist tone (see for example Lyman, 1987). Neither should it be regarded as an environment where 'piss-taking' and other forms of humour necessarily followed an entirely predictable pattern. In addition to actual focus, style of comment also varied between drawn-out story, established joke, and more credible 'one-liners'. Perhaps most surprising of all, however, was the severity with which these supposedly humorous remarks were often administered.

> [This morning conversation in the YT changing-room centred on Charlie's sexual appetite and exploits.] As we talk Charlie tells us about the girl he 'pulled' at the United reserve team match. He gives all the preliminaries then ends up saying he kissed her under the back of the stands. [Simon]: 'You'd have been fuckin' dead if someone had've caught you'. [Charlie]: "I was horny as fuck last night, I couldn't give a shit." [Simon]: "Yeah, but to be fair, she was fuckin' mingin" [not particularly attractive]. [Charlie]: "Yeah, but I wouldn't have said no to a blow-job." [Simon]: "You'd let anybody give you a blow-job." [Charlie]: "I'd let mi own mum give me a blow-job and you haven't seen her . . . fat slut." (Fieldnotes, 12.01.94)

> [YT changing-room prior to morning training session]. [Neil]: "Last night I dreamt mi' mum was up here for some reason." [Pete]: "What an' all the lads were shaggin' her." [Neil]: "Yeah, an' I was shaggin' her as well." [Charlie]: "So you were shaggin' her, an' all the lads were, and your dad was sat in the corner givin' it one'o'them [gestures masturbation] with the other finger up his arse?" [Neil]: "Yeah, he was." [Charlie]: "What an' then that baby was here that your sister's just had, and your dad was shaggin' the baby as well?" [Neil]: "That's just sick that is. There's something wrong with you. Sick bastard." (Fieldnotes, 21.01.94)

Whilst women, heterosexual exploit and masturbation were all frequent topics of daily conversation at Colby, shop-floor culture also displayed a range of more negative verbal traits around which the formation of masculine identity took place. Most apparent in this respect were inferences towards issues of homosexuality and ethnicity. In terms of the hierarchical masculine structure in

place at the club, in addition to females, individuals of minority ethnic descent
and/or homosexual tendency were vehemently regarded as inferior to the
hegemonic trainee ideals *in situ*.

In the case of ethnicity, squad relations were particularly conducive to verbal
chastisement in that all trainees were white. The most obvious manifestations
of racist behaviour by trainees took place during educational day-release at
Walton Grange College of Further Education. Here a number of first years
regularly made explicit reference to members of both Asian and Black student
groups as 'niggers', 'woggs', 'coons', 'black bastards' and/or 'Joe Daki's
(Paki's)' In this sense, boys of alternative ethnicities appeared to be regarded as
some kind of masculine threat who irritated the majority of trainees by the way
in which they adopted a casual, laid-back demeanour and communicated more
effectively with a range of female students.

> [Walton Grange College: Lunchtime]. Sat in the canteen at dinner Steve Williamson begins
> to air his thoughts on people of different ethnicities, and to assess mine. [Steve]: "I hate
> woggs, they walk round as if they own this fuckin' place. Are you racist Andy?" [AP]: "No,
> not really, I don't mind who people are or where they come from." [Steve]: "I am. I'm
> racist. I hate pakis, they think they own this fuckin' place. There's fuckin' loads of 'em
> where I come from at 'ome." (Fieldnotes, 03.02.94)

Homosexuality proved an equally problematic notion for trainees to accept.
Heterosexual standards amongst youth team members were straightforward.
Males failing to enact the basic physical and verbal masculine expectations of
footballing life – excessive drinking, sporting prowess and the vehement
pursuit of women and (heterosexual) sex – necessarily received a barrage of
criticism as regards their 'queer bastard' potential. However, in relation to both
these areas, contradictions in trainee behaviour did emerge. Despite their social
attitudes towards issues of ethnicity, for instance, trainees spoke highly of
Black college classmates and the small number of senior Black players on
contract at Colby. Rather than being regarded in a negative sense these
individuals were admired for their contemporary dress-sense, their manifestation of 'attitude' and for the commonly accepted fact that they were 'quick
fuckers' to play against. Ironically, even amidst these observations there was
evidence of dated racist assumptions concerning the supposed physical
attributes of Black sportsmen (see Cashmore, 1982; Lashley, 1980; Williams,
1994). Nonetheless, a kind of social/occupational double standard was clearly
apparent here, similar to that evident where issues of homosexuality were
concerned. Indeed, whilst being gay was completely anathema to trainee
masculine logic, discussion of national 'AIDS day' and wider sexual issues did
provide some grounds upon which individuals began to reconsider their
opinions towards such matters.

[Back at the digs we watch MTV with the latest Madonna video.] [Simon]: 'I mean how can anybody be fuckin' gay with women like that about? My mum says that they're born with women's attitudes, that's why they're like that'. We continue to discuss Channel 4's AIDS-day programmes. [Simon]: "It makes me fuckin' sick just thinkin' about it." I put it to Simon that footballers do things which other people might consider far from heterosexual. [AP]: "I mean, we all walk round naked, and get in the bath together, maybe people would think that was strange?" [Simon]: "Yeah, they might. I mean, you wouldn't go home to your mates an' talk about wankin' would you – I suppose its different when lads live together." [Gary]: "I mean they say one in five is gay!" [Simon]: "So fuckin' three of them in that changing-room is fuckin' queer! I bet Turner is, he's always touchin' your dick an' that. An' Neil, he just sits there an' stares at your prick sometimes." [Gary] "Yeah, and he's always kissin' you an' that . . ." (Fieldnotes, 15.12.93)

Such data provide clear confirmation that alongside issues of consumption and personal wealth, heterosexuality and homophobia were (on the surface at least) key aspects of masculine construction at Colby Town, and lifestyle elements crucial to the basis upon which trainee rites of passage took place. But that is not to say that trainee identities were indistinguishable. Rather, from the data presented and from my own observations in the field, three broad categories of masculine formation were generally apparent amongst youth team members. First, a 'Careerist' masculine approach could be seen to be adopted by a number of individuals who placed official institutional demand over and above issues of social importance. Second, and for the majority of trainees, a 'Conformist' masculinity was evident whereby social and official lifestyle values were accorded equal consideration. Third, and finally, a minority of boys chose to negate club stipulations and in embracing too soon the liberties of 'professional progression' attempted to live out something of a 'Bad-Boy' professional player image. In doing so these trainees not only ostracised themselves from the trainee group at times, but necessarily confirmed their lack of commitment to club values thus severely jeopardising their entitlement to managerial favour and seriously hindering their chances of occupational success.[6]

CONCLUSION

This chapter has mapped out the key issues around which gender relations and trainee masculinities were constructed at Colby Town. Within it I have pointed out that in constituting some kind of overall *rite de passage* the two year Youth Training period at the club was, in itself, a pivotal stage in the masculine development of the individuals concerned. Specifically, I have suggested that whilst various forms of youth training have traditionally represented a transitional phase between school and work, within the context of Colby Town this two year vocational programme also played host to a mid-term transition which saw all the respondents progress from trainee to 'professional' contractual conditions. This stage of the occupational maturation process, I have inferred, was a time when, as a consequence of increased monetary reward, trainee interest was diverted away from the 'official' institutional ideals of identity construction and more readily towards the social and 'unofficial' aspects of club culture.

This change of priority was primarily one of psychological and financial investment. Once trainees had both the confidence and the monetary freedom to express their personal desires they did so via lifestyle areas which they regarded as synonymous with professional player status – sexual endeavour, conspicuous consumption and excessive socializing. That these areas were considered appropriate and obligatory was implicitly linked, I have argued, to a process of 'occupational inheritance' whereby trainees carefully observed and adhered to the social habits of professionals, thus readily resisting the domestic inferiority of early institutional life and further anticipating their complete socialization into that full-time occupational role.

In short, trainees longed to be able to experience the 'manly' benefits of the professional game – to live-out the hyper-masculine practices of personal extravagance and superstar status, to enjoy the delights of fast cars, designer clothes, financial affluence, social indulgence and sexual promiscuity. In fact insofar as such images heavily infiltrated and informed most aspects of Colby Town youth team existence, they represented the epitome of trainee desire and expectation. Youth Traineeship at Colby was, quite simply, "all about becoming a man"; about graduating to professional status, about rising above the physical depression of injury, about casting aside the psychological pressures of team selection, about resisting verbal chastisement and personal humiliation, and about safely negotiating one's own masculine prowess whilst at the same time fulfilling the stringencies of club demand. Admittedly, the dual bind of trainee subordination and inferiority did have its drawbacks, but they would soon be gone; at which point the inevitability of professionalism and all its benefits would supposedly be embraced.

NOTES

1. In the interests of anonymity pseudonyms have been used throughout this chapter. The data presented are taken from Ph.D fieldwork research carried out by the author over the course of the 1993/94 footballing season. As is the trend nationally, apprenticeship within professional football is currently undergoing a period of

modification and transition whereby previously established two-year YT trainee arrangements are being superceeded by a three-year scholarship (16–19) model based on a Modern Apprenticeship framework and format. For more on the overall structure of Modern Apprenticeships in Britain see Gospel & Fuller (1998). It is worth noting, of course, that YT as a training concept was re-titled 'National Traineeship' in August 1997.

2. For further discussion concerning the adaptation of Gramsci's (1971) notion of hegemony to issues of masculinity see Carrigan et al.(1985), Parker (1992) and Jefferson (1994). For a more in-depth discussion of methodological issues within the context of masculine construction see Haywood & Mac an Ghaill (1998).

3. The managing agent for vocational education and training programmes within English professional football is the Footballers' Further Education and Vocational Training Society (FFE and VTS) which was launched in 1978 as a joint venture between the Professional Footballers' Association (PFA) and the Football League (FL) to oversee the educational and vocational needs of all PFA members. The FFE and VTS is a registered charity which is now financed jointly by the PFA, The Football League, The Football Association and The Premier League. The Football Association joined the venture in 1990, and The Premier League in 1992 (see FFE and VTS, 1993).

4. Some trainees did admit to having 'experimented with' or 'tried' substances such as 'dope' (smoking cannabis) and/or 'E' (Ecstacy) in the company of friends both during, and prior to, their time at Colby. These experiences, it was claimed, had been largely confined to isolated incidents 'at home', none of which had recurred or persisted to any significant degree. For greater insight into the problems of alcohol, drug and gambling abuse within the context of professional footballing life see Merson (1996) and Adams (1998).

5. For a more in-depth analysis of control and desexualization within organizations see Burrell (1984).

6. It is acknowledged that the adoption of such broad descriptive categories holds inherent limitations and dangers. However, it would be fair to say that the vast majority of trainees could be accurately located within one of these distinctive 'masculine' categories in terms of the occupational attitudes and demeanors they exhibited.

REFERENCES

Adams, T. (1998). *Addicted*. London: Collins/Willow.
Best, G. (1990). *The Good, The Bad, and The Bubbly*. London: Simon and Schuster.
Beynon, H. (1975). *Working for Ford*. Wakefield: EP Publishing.
Brooking, T. (1981). *Trevor Brooking*. London: Pelham Books.
Burrell, G. (1984). Sex and Organizational Analysis. *Organizational Studies*, 5(2), 97–118.
Canaan, J. E. (1996). 'One Thing Leads to Another': Drinking, Fighting and Working-Class Masculinities. In: M. Mac an Ghaill (Ed.), *Understanding Masculinities: Social Relations and Cultural Arenas* (pp. 114–125). Buckingham: Open University Press.
Carrigan, T., Connell, R. W., & Lee, J. (1985). Towards a New Sociology of Masculinity. *Theory and Society*, 5(14), 551–602.

Cashmore, E. (1982). *Black Sportsmen*. London: Routlege and Kegan Paul.

Chapman, L. (1993). *More Than a Match*. London: Arrow Books.

Clarke, J. (1973). Football Hooliganism and the Skinheads. Occasional Paper. CCCS, University of Birmingham. January.

Clarke, J. (1976). Style. In: S. Hall & T. Jefferson (Eds), *Resistance Through Rituals* (pp. 175–191). London: Hutchinson.

Clarke, J. (1978). Football and Working Class Fans: Tradition and Change. In: R. Ingham, (Ed.) *Football Hooliganism* (pp. 37–60). London: Inter-Action.

Clarke, J. (1979). Capital and Culture: the post-war working-class re-visited. In: J. Clarke, C. Critcher & R. Johnson (Eds), *Working Class Culture* (pp. 238–253). London: Hutchinson.

Cockburn, C. (1983). *Brothers*. London: Pluto Press.

Cockburn, C. (1987). *Two-Track Training*. London: Macmillan.

Collinson, D. L. (1988). Engineering Humour: Masculinity, Joking and Conflict in Shop Floor Relations. *Organization Studies*, *9*(2), 181–199.

Collinson, D. L. (1992). *Managing the Shop Floor*. London: DeGruyter.

Connell, R. W. (1987). *Gender and Power*. Cambridge: Polity Press.

Cowie, C., & Lees, S. (1981). Slags or Drags. *Feminist Review*, *9*, 17–31.

Critcher, C. (1972). Football and Cultural Values. Working Papers in Cultural Studies no.1. University of Birmingham: 103–119.

Douglas, P. (1973). *The Football Industry*. London: Allen and Unwin.

Featherstone, M. (1991). The Body in Consumer Culture. In: M. Featherstone, M. Hepworth & B. S. Turner (Eds), *The Body: Social Processes and Cultural Theory* (pp. 171–197). London: Sage.

Footballers' Further Education and Vocational Training Society. (1993). *Football League YT Training Proposal*. Manchester: PFA/FFE and VTS.

Giulianotti, R. (1993). Soccer Casuals and Football Intermediaries. In: S. Redhead (Ed.), *The Passion and the Fashion: Football Fandom in the New Europe* (pp. 153–205). Aldershot: Avebury.

Gospel, H., & Fuller, A. (1998). The Modern Apprenticeship: new wine in old bottles? *Human Resource Management Journal*, *8*(1), 5–22.

Gowling, A. (1974). The Occupation of the Professional Footballer. Unpublished MA Thesis, Victoria University of Manchester.

Gramsci, A. (1971). *Selections from the Prison Notebooks*. London: Lawrence and Wishart.

Gray, S. (1987). Sharing the Shop Floor. In: M. Kaufman (Ed.), *Beyond Patriarchy* (pp. 216–234). Toronto: Oxford University Press.

Griffin, C. (1993). *Representations Of Youth*. Cambridge: Polity Press.

Hall, S., & Jefferson, T. (1976). *Resistance Through Rituals*. London: Hutchinson.

Hargreaves, J. (1987). *Sport, Power and Culture*. Cambridge: Polity Press.

Haywood, C., & Mac an Ghaill, M. (1998). The Making of Men: Theorizing Methodology in 'Uncertain Times'. In: G. Walford (Ed.), *Doing Research about Education* (pp. 125–138). London: Falmer Press.

Hendry, L. B. et al. (1993). *Young People's Leisure and Lifestyles*. London: Routledge.

Hollands, R. G. (1990). *The Long Transition*. London: Macmillan.

Hollway, W. (1984). Gender Difference and the Production of Subjectivity. In: J. Henriques et al. (Eds), *Changing the Subject* (pp. 227–263). London: Methuen.

Hopcraft, A. (1971). *The Football Man*. Harmondsworth: Penguin.

Jefferson, T. (1994). Theorizing Masculine Subjectivity. In: T. Newburn & E. A. Stanko (Ed.), *Just Boys Doing Business* (pp. 10–31). London: Routledge.

64   ANDREW PARKER

Jenkins, R. (1983). *Lads, Citizens and Ordinary Kids.* London: Routledge.
Lashley, H. (1980). The New Black Magic. *British Journal of Physical Education, 11*(1), 5–6.
Lyman, P. (1987). The Fraternal Bond as a Joking Relationship: A Case Study of the Role of Sexist Jokes in Male Group Bonding. In: M. S. Kimmel (Ed.), *Changing Men* (pp. 148–163). London: Sage.
Mac an Ghaill, M. (1994). *The Making of Men: masculinities, sexualities and schooling.* Buckingham: Open University Press.
Merson, P. (1996). *Rock Bottom.* London: Bloomsbury.
Parker, A. (1992). One of the Boys? Images of Masculinity Within Boys' Physical Education. Unpublished MA Thesis. University of Warwick, Coventry, U.K.
Parker, A. (1996a). Chasing the Big-Time: Football Apprenticeship in the 1990s. Unpublished Ph. D.Thesis. University of Warwick, Coventry, U.K.
Parker, A. (1996b). Sporting Masculinities: Gender Relations and the Body. In: M. Mac an Ghaill (Ed.), *Understanding Masculinities: Social Relations and Cultural Arenas* (pp. 126–138). Buckingham: Open University Press.
Parker, A. (1998). Staying On-Side on the Inside: Problems and Dilemmas in Ethnography. *Sociology Review, 7*(3), 10–13.
Parker, H. (1974). *View From The Boys.* London: David and Charles.
Redhead, S. (1991). *Football With Attitude.* Manchester: Wordsmith.
Riemer, J. W. (1979). *Hard Hats.* London: Sage.
Sabo, D. F., & Panepinto, J. (1990). Football Ritual and the Social Reproduction of Masculinity. In: M. Messner & D. F. Sabo (Eds), *Sport, Men and the Gender Order* (pp. 115–126). Champaign, Illinois: Human Kinetics.
Shankly, B. (1977). *Shankly.* London: Book Club Associates.
Taylor, I. (1971a). Soccer Conciousness and Soccer Hooliganism. In: S. Cohen (Ed.), *Images of Deviance* (pp. 134–164). Harmondsworth: Penguin.
Taylor, I. (1971b). Football Mad. In: E. Dunning (Ed.), *The Sociology of Sport*, London: Frank Cass.
Taylor, R., & Ward, A. (1995). *Kicking and Screaming: An Oral History of Football in England.* London: Robson Books.
Tolson, A. (1977). *The Limits of Masculinity.* London: Tavistock.
Walker, J. C. (1988). *Louts and Legends.* Sydney: Allen and Unwin.
White, (1971). Rituals of the Trade. *New Society*, 13 May: 805–807.
Williams, J. (1994). 'Rangers is a Black Club': Race, Identity and Local Football in England. In: R. Giulianotti & J. Williams (Ed.), *Game Without Frontiers: Football, Identity and Modernity* (pp. 153–183). Aldershot: Arena.
Williams, J., & Taylor, R. (1994). Boys Keep Swinging: Masculinity and Football Culture in England. In: T. Newburn & E. A. Stanko (Ed.), *Just Boys Doing Business* (pp. 214–233). London: Routledge.
Williams, J., & Woodhouse, J. (1991). Can Play, Will Play? Women and Football in Britain. In: J. Williams & S. Wagg (Ed.), *British Football and Social Change* (pp. 85–108). Leicester: Leicester University Press.
Willis, P. E. (1977). *Learning To Labour.* Farnborough: Saxon House.
Willis, P. E. (1979). Shop Floor Culture, Masculinity and the Wage Form. In: J. Clarke, C. Critcher & R. Johnson (Eds), *Working Class Culture* (pp. 185–198). London: Hutchinson.
Willis, P. E. (1990). *Common Culture.* Milton Keynes: Open University Press.
Willmott, P. (1966). *Adolescent Boys of East London.* London: Routledge and Kegan Paul.

Wilshaw, D. (1970). A Sociological and Psychological Enquiry into Association Football, with Specific Reference to Vocational Guidance. Unpublished M.Ed. Thesis. University of Manchester, Manchester, U.K.

Wood, J. (1984). Groping Towards Sexism: boys sex talk. In: A. McRobbie an& M. Nava (Ed.), *Gender and Generation* (pp. 54–84). London: Macmillan.

FOOTBALL AND RELATIONSHIPS: GENDERED EXPERIENCES OF HOME?

Caroline Hudson

> She wanted to discuss our relationship. I wanted to watch the football, so she threw the telly at me.

At a conscious level, I had forgotten the comment above, made to me some years ago by a male friend, until analyzing and writing up the data from my doctoral ethnography. However, the comment's terse suggestion of potential gender differences in perceptions of relationships has resonated with some of the voices from my doctoral ethnography. Together, they raise questions which are explored in this chapter.

THE DOCTORAL ETHNOGRAPHY

The doctoral ethnography (Hudson, 1999) examined the extent to which 32 young people perceived that their family structure (intact nuclear, reordered nuclear or step, and single-parent) influenced their experience of family and schooling. Through attempting to listen to the young people's voices on issues related to home and school, the study questions whether commonly used categories of family structure (intact nuclear, reordered nuclear or step, and single-parent) inform understanding of these young people's experience of family and schooling.

The ethnography was of one mixed ability tutor group of 32 students in a large urban comprehensive school in central England. Fieldwork was conducted over two years, from September 1996 to July 1998, though the main period of data collection was from November 1996 to July 1997, when the

Genders and Sexualities in Educational Ethnography, Volume 3, pages 67–89.

young people were in Year 9. The main methods of data construction were interviews and informal conversations with the young people, and observation of their experience of schooling. Two sets of tape-recorded semi-structured interviews were conducted. The first of these was in January–February, 1997 and the second of these was in April–May, 1997. Additional tape-recorded interviews were conducted with those students who requested them. In all, 237 tape-recorded interviews were conducted.

THE RESEARCH LITERATURE ON FAMILY STRUCTURE AND GENDER

There is a vast body of research literature on family structure. Most of this literature is quantitative, focusing on young people's outcomes relating to health, emotion and behaviour, and education, according to young people's family structure (intact nuclear, reordered nuclear or step, and single-parent). On the whole, the literature argues that whilst there is no path down which children will inevitably travel, children who have experienced the break-up of their parents' relationship tend to run greater risks of poorer health, emotional, behavioural and/or educational outcomes than their peers in intact nuclear families (Burghes, 1994; Kiernan, 1998; Utting, 1995). The outcomes are, however, averages for the groups under study, and conceal the best and worst outcomes within each group. The differences between children living in intact nuclear families and those in other family structures are often small (Cockett & Tripp, 1994; Rodger & Pryor, 1998; Utting, 1995) and disadvantaged outcomes identified by research apply only to a small minority of children. There is, furthermore, a range of methodological problems with the quantitative data on family structure.

There exists little qualitative British research on the processes leading to young people's outcomes, according to family structure. Most existing qualitative research explores adults' responses to family reordering, or adults' perceptions of children's responses to family reordering. Few British qualitative studies on family structure aim to access the voice of the child; the chief exceptions to this are Walczak & Burns (1984), Mitchell (1985) and Cockett & Tripp (1994).

Whilst gender is a variable frequently used in quantitative analyses of children's outcomes according to family structure, it is not within the scope of the quantitative literature to tease out in detail the potential influences of gender in children's responses to family reordering. Within the qualitative literature on family structure which purports to listen to the voice of the child, the voices are usually strikingly ungendered. For instance, Mitchell (1985, p. vii) claims that she is exploring "children's feelings and experiences in their own words" about family separation and divorce, but she describes her sample in ungendered terms as: ". . . young people who had been under the age of sixteen" (p. 13).

On the whole, Mitchell discusses her findings as though 'feelings and experiences' are unproblematically comparable across the genders. Even on the occasions when she does make distinctions about responses to family separation according to gender, as when saying that more girls than boys said that they had talked to someone about the family break-up, potential gender differences are not fully teased out. The 'Discussion and Recommendation' chapter (pp. 177–191) does not refer to gender – instead, she uses the undifferentiated word 'children'.

Utting (1995) comments of *The Exeter Family Study* (Cockett & Tripp 1994) that, "The research was unusual in the efforts made to 'listen to children'" (p. 48).

Although Cocotte & Tripp (1994) point out that girls were more likely than boys to have talked to a sibling about family separation, the study elsewhere refers to 'children' with the implicit assumptions that, within Cockett & Tripp's (1994) sample, first, boys' and girls' experiences of family were similar, and second, boys and girls presented their experiences of family in similar ways. Cockett and Tripp's near-omission of any consideration of gender in their data set is particularly striking in that, when discussing previous research on family structure, Cockett and Tripp do summarize the influences of gender in children's responses to family reordering.

The silences, or near-silences, on gender by researchers on family structure, such as Cockett & Tripp (1994), speak loudly, given the co-existence of an active academic debate on gender over recent decades. Some feminists (e.g. Deaux & Major, 1990; Thorne, 1990) argue that much academic writing on gender has tended to highlight differences between genders, rather than potential similarities across the genders. For instance, researchers such as Deaux & Major (1990), Maltz & Borkner (1998), and Tannen (1998) argue that women have often been presented as expressive, empathetic, and concerned with interpersonal relationships and connection with others; in contrast, men have often been presented as more instrumental and assertive than women. Thorne (1990) claims that research often highlights the cooperative nature of girls', in contrast to boys', social relations. Gilligan (1982), in her controversial book on how men and women make moral decisions, argues that women speak "in a different voice" to men; according to Gilligan, men tend to use abstract moral principles to make moral decisions, whereas women's moral decisions tend to be shaped by love and empathy for the individuals concerned.

Researchers such as Thorne (1990) and Deaux & Major (1990) argue that simplistic gender dualisms potentially obscure the complexity and fluidity of gender relations. Deaux & Major (1990) point out that not only may some individuals be more aware than others of their gendered identity, but also gender is likely to shape experience more on some occasions, such as when a woman gives birth, than on others. Thorne (1990) underlines that acknowledging moments at which gender appears irrelevant, as well as occasions on which gender is important, is key to deepening understanding of power dynamics in social relations. Freedman (1990), in turn, argues that what is needed in analyses of gender is a theoretical framework which is flexible enough to reflect both difference and similarity.

Though the focus of the ethnography was family structure, not gender, the influences of gender were apparent in its findings, both methodological and substantive. This chapter will, first, illustrate how potential gender imbalances in field relations between researcher and researched were addressed. Second, the chapter will explore the extent to which the influences of gender were apparent in research subjects' accounts of family relationships, to assess how far consideration of gender potentially informs academic research on family structure. The analysis aims to avoid simplistic dichotomies about gender, as discussed above; it, rather, attempts to recognize both differences between and similarities across the genders.

GENDER AND FIELD RELATIONS

Within the literature on research methodology, it has become almost a platitude that gender is likely to impact upon the relationship between researcher and researched. It is probably fair to say that, in most circumstances, female researchers are likely to elicit greater openness from the research subject than male researchers, particularly when issues discussed are especially sensitive, when the research subject is female, and when a non-hierarchical stance is adopted in the research relationship (Finch, 1984). It could therefore be argued that, because of my gender, I was likely to construct franker data from research subjects than a male researcher would have, and that girls would probably be more open with me than boys.

However, Silverman (1993, p. 35), for instance, emphasizes the importance of researcher reflection on the influences of gender on the data constructed. With implicit similarities to, for instance, Thorne (1990) and Deaux & Major (1990), as discussed above, Silverman warns against universal, taken-for-granted assumptions about gender which may, in actuality, be culturally and historically specific. Silverman (1993, p. 35) cites McKeganey & Bloor's

(1991, pp. 195–6) view that other factors, such as social class and age, may impact upon fieldwork, and that the influences of gender may not be ascribed, but negotiated, with the researched.

My chief aim in negotiating field relations was to reduce power relations between researcher and researched, by giving the young people as many choices as possible, including: whether to be interviewed; when to be interviewed; how often to be interviewed, and how much to say in interview. Within this broad aim, I attempted to minimize potential gender imbalances between researcher and researched. Because I was in the field for blocks of time over a considerable period, I became aware of different ways of building field relations with the boys and girls in my sample. Acting on this awareness helped me to develop relationships of trust with both boys and girls. It is likely that these relationships, in turn, encouraged both boys and girls to speak openly, in interview or in informal conversation with me, about experiences of family and schooling.

It may appear paradoxical that developing shared jokes with boys was probably a key preliminary stage in enabling boys to want to talk to me on a one-to-one basis subsequently. A large friendship group of boys developed two central jokes with me, which they replayed time and time again during early stages of fieldwork. One morning, two boys noticed I was wearing two watches (Research Log 27.11.96). They not only questioned my reasons for this, but also enjoyed asking me, on many subsequent occasions, if I had two watches on. The second joke was about cars. In a Health, Home and Community (HHC) lesson, I told this group of boys that I had bought a new car (Research Log, 28.11.96). One of the boys, Matty, suggested that his father could have sold me one of his four cars cheaply. I suggested it was not too late to ask Matty's father. In subsequent conversations, the boys returned, over and over again, to the subject of whether I was going to buy Matty's father's car. One boy, Johnny, would even remind me to ask Matty about his father's car (Research Log, e.g. 5.12.96, 12.12.96). Another student, Muhammed, subsequently joined in the dealings, by asking his father if he would sell me a BMW from his taxi business. Muhammed seemed to enjoy trying to strike a deal with me during a number of lessons.

In contrast, the development of my relationships with girls took a different pattern. One girl, Sara, adopted the role of key student gatekeeper almost as soon as fieldwork began. In the first tutor time I spent with the young people, I explained that I was writing a book on Year 9 students' views on family and schooling, and that I wanted the students to act 'normally' and 'naturally' with me. Sara caught on to this immediately, and she not only appeared to need no encouragement to open up to me herself, but she would also frequently remind her peers that I wanted them to, in Sara's words, 'act normal' or 'act natural'. For example, when Anna, one of Sara's friends, heard Sara telling me how the two of them had almost been caught smoking in the school ditch at lunchtime, she said, alarm in her voice, "You haven't told her that, have you?" Sara responded, "She wants us to act natural"; Anna joined in the conversation. Because it was Sara who initially would invite me to sit with her and her friends in class, from near the beginning of fieldwork I had access to girls' peer group conversations. Soon, other girls started taking the initiative in inviting me to sit with them in class.

Whilst Sara undoubtedly facilitated my field relations with other girls, the girls, in contrast to the boys, did not need to develop shared jokes with me as part of relationship-building. Though sharing jokes and talking intimately to someone might at first glance appear at odds with one another, I speculate that the jokes the boys shared with me over a period of time helped create the context in which they could talk with me on a one-to-one basis about family relationships. In contrast to the development of field relations with boys, the growth in the closeness of my relationships with girls was, with Sara's help, much more cumulative, from the beginning of fieldwork onwards. It is likely that sensitivity to different patterns in the development of relationships with boys and girls helped minimize potential gender imbalances in the substantive findings of the doctorate.

GENDER DIFFERENCES AND FAMILY RELATIONSHIPS

Relationships with Mothers

All the young people except three (Stacey and, for part of the fieldwork, Brian and Martin) lived with their natural mothers. However, it does not follow from this structural similarity in living arrangements that the young people would necessarily experience similar relationships with mothers, across the family structures. It could be argued, for example, that young people's relationships with mothers might vary according to family structure, because the interpersonal dynamics within the immediate family in reordered families might differ greatly from those of intact nuclear families. It could be claimed that potential differences in family dynamics, according to family structure, might impact upon young people's relationships with mothers.

Across the range of family structures, the young people's accounts suggested that they valued communication in their relationships with their mothers. However, on the whole, the girls' accounts tended to highlight a greater degree of closeness between themselves and their mothers than was revealed in the boys' accounts of relationships with mothers.

For example, as Rich (1990) found in her research with girls at Emma Willard School, Clare (intact nuclear) described how she confided in her mother; she told her mother 'secrets' about herself (Interview, December 1997). Clare, Anna (reordered nuclear) and Fiona (reordered nuclear) explicitly stated that they could talk to their mothers, but not to their fathers, whether natural or step, about relationships with boys. In various interviews, Rebecca (intact nuclear) emphasized how she and her siblings were close to their mother, in contrast to their father.

Louise particularly valued the period during fieldwork when her step-father had left and she was in a single-parent, mother-headed household. This was partly because Louise greatly enjoyed what she perceived as the adult relationship she shared with her mother in which, according to Louise, they often talked together intimately. What follows is an extract from Louise's account of a much longer conversation with her mother, in which they comment derogatorily on men:

> I know more about men than most people, 'cos my mum talks to me about men, She says, "What are men?" I say, I'm allowed to say this word about men, 'W-a-n-k-e-r-s' (spells it out). And she asks me, "What do you want 'em for?" And I go, "Their money?" An' she goes, 'Yep.' And she tells me like how to treat men. You're meant to treat men like they're gonna treat you . . . (Interview, April 1997).

Like Louise, Leanne said that she had valued the time she and her older sister had spent in a single-parent, mother-headed family, after her natural father had left and before her mother had remarried. Leanne not only highlighted how the three of them had been very close during this period, but she also stressed how their intimacy had provided the basis for the close relationship she had subsequently enjoyed with both her mother and her sister:

> Being close in a family is really important, 'cos I can speak very easily to my mum. She's really the only person I can really talk to and my older sister, because when I was younger it was just the three of us for quite a few years, so I can speak very openly to my mum (January, 1998).

Across all family structures, whilst boys valued communication in their relationships with their mothers, their relationships appeared, on the whole, less close than girls' relationships with their mothers. Some boys highlighted the importance of conversation in their relationships with mothers. Matty (intact nuclear) said he talked more to his mother than to his father (Interview, May, 1997), whilst Malcolm (intact nuclear) described how:

> When my mum gets home, we normally have a fag together, and talk over the day (Interview, April, 1997).

Two boys explicitly talked about how an important aspect of their relationships with their mothers was that their mothers were understanding. Steven (intact nuclear) said that his mother was more understanding than his father and that she was becoming increasingly so (Interview, April, 1997). Charley (reordered nuclear) focused explicitly on his mother's awareness of his feelings:

> My mum knows when I'm emotional and how to comfort me (Interview, April, 1997).

The majority of boys, however, explored the degree of closeness and communication in their relationships with their mothers in less detail than most of the girls.

To some extent, family structure and gender interacted in the young people's presentations of relationships with mothers. Whatever the family structure, the girls in the sample tended to show greater awareness of their mothers' emotions and needs (Rich 1990). Within this sample, however, it was some young people in reordered families (Leanne, David, Charley, Tim, Anna, Fiona, Martin, Louise, Sara) who showed particular awareness of their resident mothers' needs and feelings. Tim, David and Charley will be discussed in the final section of this chapter. Of the girls listed above, Leanne, for example, revealed how she fulfilled her mother's need for closeness, which she related to her mother's sense of security:

> My mum likes that I speak to her, because then she feels a part of me. Like if I didn't speak to her, I don't think she would feel that happy, because she wouldn't feel secure enough (Interview, April, 1997).

Whilst Sara's (single-parent) relationship with her mother was often turbulent, and whilst Sara on occasions explicitly highlighted her emotional distance from her mother, at the same time she occasionally also showed considerable sensitivity towards her mother's feelings. For example, some weeks in advance, Sara bought a cafetiere for her mother for Mother's Day, because her mother apparently had wanted one for a long time. Sara had even planned out how she was going to give her mother the cafetiere, to maximize her mother's pleasure:

> Cos I've got 'er, 'cos she's wanted for ages, she's always wanted, you know, one of them coffee plunger things. You know, where you press the thing down. Well, I got that. And I thought, "There's no point just gettin' that without any coffee, 'cos it's gonna be on a Sunday," so I got her some of that as well. An' I'll just take that up, to see 'er face. I'll say, "Well, I'll get you a cup of coffee then, and then I'll give you your present." So I wrapped the jar, 'cos it was in a packet, so I thought, "I'll put it into a jar." And I put some in another jar, so I could take it downstairs and use it (Interview, February, 1997).

However, it should not be assumed that awareness of their mothers as people with needs and feelings meant that these young people would necessarily attempt to meet these needs on all occasions. For example, Sara's father had died two years before fieldwork started. Sara talked in detail over a number of interviews about how, to preserve her own emotional composure, she consciously distanced herself from her mother's grief over his death, as the following indicates:

> When my dad died my mum was like really upset and that an' I just used to walk out of the room all the time, 'cos I hated, if I was thinkin' about my dad, I just like used to go and sit on my own. I still do it. 'Cos I just can't handle it when my mum starts cryin' (Interview, February 1997).

Anna, in turn, refused to be drawn into her mother's problems with her non-resident step-mother:

> You know, if my mum wants to hate Gertie [Anna's non-resident step-mother], then that's her problem, but I'm not gonna get involved and I'm not gonna take sides (Interview, February, 1997).

In the young people's accounts of relationships with mothers, the influences of gender potentially inform understanding of family structure. Both the similarities in the young people's relationships with their mothers, across family structures, and the greater closeness highlighted by girls, in comparison to boys, in accounts of relationships with mothers, underline the importance of gender in the young people's relationships with mothers. In turn, these points suggest that family structure was largely unimportant in these young people's relationships with mothers. At the same time, there was an association between family reordering and greater awareness of mothers' needs and feelings. This may be related to claims made by a minority of researchers on family structure, such as Walczak & Burns (1984), that, through the experience of family reordering, some research subjects claimed to have developed a better understanding of people and of human relationships. It could be argued that awareness of others' needs and feelings, as manifested by some of these young people in reordered families, is a positive indication of maturity. This contrast with the research literature's emphasis on the negative aspects of family reordering underlines the perhaps excessively deficit model of family reordering in much existing research literature.

Relationships with Fathers

In most reordered family networks, the natural father is non-resident. The literature on family structure tends to argue that children in reordered families are likely to experience unsatisfactory relationships with the non-resident natural father. It is claimed, for instance, that contact with the non-resident natural father is, in many cases, infrequent and, furthermore, lessens over time. Burghes (1994) highlights the need for more research on how children's relationships with their fathers change before and after separation. However, it is only a minority of researchers on family structure (e.g. Furstenberg & Cherlin, 1991) who point out that many children in intact nuclear families may not get the attention they want from their fathers, and that it is not known if fathers in intact nuclear families are more involved in their children's upbringing than non-resident natural fathers of children in reordered families. Instead, the literature on family structure tends to start from the unarticulated assumption that there is a qualitative difference between the relationships of young people in intact nuclear and in reordered families with their fathers, natural resident, natural non-resident or step. Consequently, the literature on family structure potentially underplays similarities in the relationships of fathers and their children, across the range of family structures. This section will explore the influences of gender in these young people's accounts of their relationships with their fathers, whether natural resident, natural non-resident or step, to assess the extent to which these relationships differ according to family structure.

Girls and boys in this sample tended to present their relationships with their fathers as more distant than their relationships with their mothers. Like the boys in Morrow's (1998) sample, four boys in intact nuclear families who presented their relationships with their natural resident fathers in positive terms, highlighted activities they shared with their fathers. Andrew, for instance, spoke of how he did the weekly shop with his father and of how, particularly on holiday, he enjoyed going to the pub with his father. Andrew apparently also gave his father moral support in his father's attempts to give up smoking (Interview, January, 1997). Several boys in intact nuclear families, such as Andrew, Muhammed and Matty, spoke enthusiastically about visiting their fathers at work and about helping them with their jobs. Tom, in turn, was in a county football team, and was enthusiastic about how his father took him to football training and watched his football matches (Interview, January, 1997).

Whilst Clare (intact nuclear) did not talk about her father a great deal in her interviews, her accounts suggested they had an amicable relationship:

The best thing about holidays is having a laugh with my dad (Interview, July, 1997).

However, Clare also explicitly commented that there was a lot about her life that she did not tell her father (Interview, April, 1997). Whilst Julie was in an intact nuclear family during fieldwork, she had experienced family reordering temporarily, prior to fieldwork, when her natural father had left home for a year.

Julie's account of her father suggested that she did not consider that their relationship was close:

> I don't really get along with my dad now, but it's got nothing to do with him leaving. It's just that he usually just gives me lifts places, and that's about it (Interview, April, 1997).

The accounts of Steven (intact nuclear) and four girls – Carla (intact nuclear), Clare (intact nuclear), Karen (intact nuclear) and Stacey (reordered nuclear) – of their resident natural fathers emphasized that their fathers spent a lot of time out at work. Steven for example, said of his father, who worked as an accountant, but who was training to be a vicar:

> My dad's usually just working, or at his religion (Interview, April, 1997).

Steven, however, appeared unperturbed by his father's absorption in work and did not develop the subject further. According to some girls, their fathers could be grumpy when they were at home. For example, Karen and Clare highlighted their perception that their fathers were 'stressy' when they were working nights. Stacey, in turn, presented her resident natural father as uncommunicative:

> My dad's not very easy to talk to. He's usually watching the football or something (Interview, April, 1997).

However, although young people in intact nuclear families tended to present their relationships with resident natural fathers as less close than relationships with resident natural mothers, and although some girls explicitly highlighted their perception of a lack of communication with their resident natural fathers, this did not appear to distress these young people.

Of the accounts of relationships with resident natural fathers, Rebecca's was the most hostile. Rebecca frequently described how she wanted her father to leave home, as the following illustrates:

> I'd like it if my dad left, 'cos we can do stuff better when he's not around (Interview, May, 1997).

It is likely that there were links between Rebecca's negative response to her resident natural father, and her father's physical violence towards her mother and his emotional violence towards Rebecca, her mother and her siblings.

There was considerable variation in the accounts of young people in reordered families, whether reordered nuclear or single-parent, of their relationships with their non-resident natural father. Anna's (reordered nuclear), David's (reordered nuclear) and Colin's (single-parent) relationships with their natural non-resident father appeared most positive. These young people had regular contact with their non-resident natural fathers, who lived locally. Anna saw her father every Sunday, Colin saw his natural non-resident father every weekend, whilst David saw his father three times a week. In their accounts of their relationships with their natural non-resident fathers, these young people not only pointed to their regular contact with their fathers, but also to activities they shared with their fathers. Anna, for instance, described how she went cycling with her natural non-resident father and her non-resident step-mother, and how she also went on holidays with them. Colin spoke of regular weekend outings to, for instance, fun fairs, and of holidays with his non-resident natural father and his father's girlfriend. David, in turn, spoke of enjoying outings with his non-resident natural father to a favourite local restaurant. This implicit link between shared activities and perceptions of positive relationships was, therefore, common to the relationships of some young people in intact nuclear families and some young people in reordered families with their natural fathers, both resident and non-resident.

Like Anna's, David's and Colin's natural non-resident fathers, Charley's (reordered nuclear) and Tim's natural non-resident fathers lived locally. However, in contrast to Anna, Colin and David, but like some of the young people in Cockett and Tripp's (1994) study, Charley and Tim only had occasional, accidental contact with their natural non-resident father. The casual nature of Charley's contact with his natural non-resident father is captured by the following account:

> Last time I saw 'im [natural non-resident father] was . . . well, ages ago. My auntie . . . needed some petrol one day, so we stopped at this garage, and 'e was there. . . 'E just asked me 'ow my mum was and I said, 'Alright', and . . . that was all that 'e said really (Interview, January, 1997).

According to Charley, the amount of contact he had with his natural non-resident father had diminished over time. Charley described how, immediately after his father had left, they used to meet several times a week (Interview, April, 1997). However, whilst there may have been a discrepancy between what Charley said and what he felt, he never indicated any wish for more regular contact with his natural non-resident father.

In Tim's first interview (January, 1997), he said that he could only remember seeing his natural non-resident father on a few occasions over the course of his life. Tim stressed his surprise and also the difficulty he experienced in talking to his father, on the first occasion he remembered meeting him:

> I couldn't believe he was my dad at first. I thought 'e was one of my mum's friends. I couldn't really talk to 'im.

Like Charley, Tim never expressed any wish to have more direct contact with his father. Indeed, Tim's mother apparently used the idea of Tim living with his natural father as a threat, when Tim misbehaved:

Tim: I'm grounded. My mum's threatened that I'll live with my dad.

CH: Does your mum mean it?

Tim: I ain't got a clue . . . This time I think she means it (in a world-weary voice).

In a later interview, Tim revealed that there were unresolved issues related to his father to which he wanted answers. In the following, Tim's father had apparently visited Tim's home two weekends before this interview:

I asked 'im [Tim's father], "Why weren't you there?" My mum hadn't said why he walked out. Said 'e 'ad too many kids anyway . . . Tough if my dad mined me asking – it was a fair question. I'd 'ave liked to ask 'im why 'e came back (Interview, April, 1997).

Leanne, Fiona and Louise described their relationships with their natural non-resident fathers in explicitly negative terms. As in Mitchell's (1985) study, both Leanne and Fiona were dissatisfied with the amount of contact they had with their natural non-resident fathers, though the frequency of the two girls' contact with their fathers differed. Leanne's father lived in Castletown, and she saw him once every two weeks, whilst Fiona's father lived in Wales, and she saw him much more intermittently. Whilst Fiona described how her father had, on various occasions, let her and her sister down about promised visits (Interview, April, 1997), Leanne described how her father had reduced the amount of contact time, from once a week to once every two weeks. Leanne explored in detail her responses to this:

It used to be once a week that I saw my dad, but then he decided we would only see each other once a fortnight. His reason was that we need a bit more space. I thought we had enough space. I wanted really to see him a bit more. He spoke to my mum and my older sister about it and I was sort of pushed to the side and told later on. I think he should have listened to us all together and listened to every one and come to a mutual decision. I said to him when I went to see him last that I thought that maybe we should get to see each other in the week a bit more, but I didn't really get a response (laughs nervously). He didn't really say anything. He didn't say anything at all (with great expression), so I felt a bit upset. He didn't realize because I didn't tell him how I felt. I feel I don't know him enough to tell him how I feel, and he never tells me how he feels (Interview, April, 1997).

The above underlines Leanne's natural non-resident father's apparent unresponsiveness to Leanne's wishes over contact time; Walczak & Burns (1984), in contrast, highlight the importance of the child having a say in access arrangements. The above also reveals Leanne's perception of her inability to communicate with her natural non-resident father. This was an issue to which Leanne frequently returned over the course of her interviews with me. For instance, near the beginning of her first interview, Leanne said that:

I don't really speak much to my dad 'cos he like left us when I was young and I haven't been able to speak very well to him (Interview, January, 1997).

In her second interview, Leanne returned to this subject:

> I can't really speak to my dad (January, 1997).

Fiona, in turn, described how her ears "sort of block(ed) up" when she and her father were talking.

Both Leanne and Fiona did not feel that their need for love was met by their natural non-resident fathers, as is encapsulated by the following:

> Leanne: (in conversation about her natural father). I feel I need a touch more loving . . . just to be able to be with him a bit more, just me and my dad. I think like 'cos I haven't sort of cuddled him for like a long, long time, and I feel that if we had a bit of a cuddle sometimes, I'd maybe feel a bit happier about our relationship.
>
> CH: When did you last have a cuddle?
>
> Leanne; I honestly can't remember (in a regretful tone). Maybe about a year ago (Interview, April, 1997).

Fiona, in turn, was regretful that:

> My real dad has never told me he loves me.

Of all the young people, however, Louise's accounts conveyed the most negative feelings towards a natural non-resident father, as is captured by the antagonism of the following:

> We used to 'ave to go and see my old dad every Saturday, an' I got in the car with 'im, an' I just wouldn't go with 'im. Something said, "No, don't go." An' he had gave me this present for my birthday an' I 'ated 'im so much I took it home an' put it in the bin . . . It was a pair of earrings 'an it said like, "To my daughter," an' I said, "No" (Interview, January, 1997).

The intensity of Louise's feelings is underlined by how she both threw away his birthday present to her and explicitly denied any father–daughter relationship. According to Louise, at the time of fieldwork, she had no contact with her natural father, and she explicitly said that she never wanted to see him again (Interview, January 1997). It is likely that Louise's feelings were related to her father's previous history of violence, when he lived with the family:

> Louise: 'E [her natural father] used to come 'ome [from the pub[and 'e was so drunk.
>
> CH: Can you remember it?
>
> Louise: Yeh, it was 'orrible. 'E used to, 'cos my brother was one years old, and 'e used to like . . . really 'it 'im . . . really bad. 'E chucked me down the stairs when I was little, and 'e put my mum's 'ead through a window. That was when my mum said, "He's gotta go" (Interview, January, 1997).

Indeed, Louise explicitly said:

I dunno whether I'd want to see my real dad again, because of what he done. I will never forgive him for that (Interview, January 1997).

Brian was in a children's home for much of the fieldwork. His account of his feelings towards his non-resident natural father is not dissimilar to that of Louise, as illustrated above. Common to both young people was their experience of their fathers' violence. Brian declared unequivocally:

I hate my dad. He abused me physically. I don't know where he is and I hope he isn't alive.

In turn, Brian's and Louise's negative feelings towards their natural non-resident fathers, were, though more extreme than those expressed by Rebecca about her natural resident father, not entirely dissimilar to Rebecca's response. This further underlines the potential influence of a father's violence upon these young people's feelings towards their fathers.

Within the sample, there was also variation in the accounts of young people in reordered nuclear families of their relationships with resident step-fathers. Leanne, Anna and David presented their relationships with their resident step-fathers in positive terms, and their relationships tended to be based on shared activities. David appeared particularly keen on the regular games of snooker he shared with his step-father (Interview, January 1997). David also emphasized how his step-father was directly involved in other activities with David:

My step-dad takes me to football trainin' an' helps me when I have to get things for teams and stuff (Interview, April, 1997).

Anna, in turn, enjoyed going sailing regularly with her step-father. In describing a sailing holiday the family shared together, Anna emphasized her mother's fear of sailing. In Anna's account of this, both her pride at managing the boat with her step-father, and her sense of togetherness in doing this with him, stood out (Interview, February, 1997).

Leanne enjoyed her weekly games of squash with her resident step-father:

I'm on my own with my stepdad, 'cos we play squash together. Like that we get to talk about things I don't really get to talk about with anyone, with my real dad. Like with my real dad there's always someone else there (Interview, January, 1997).

The above reveals, first, that Leanne equated shared activities with her resident step-father with spending what appears to be valued private time with him, and second, that Leanne viewed her relationship with her resident step-father as closer than that with her non-resident natural father. Indeed, Leanne was explicit about the latter on other occasions; in a later interview in the same month as the above, for instance, she declared categorically:

I know him [her step-father] better than my real dad (Interview, January, 1997).

Leanne's accounts highlighted her resident step-father's affection towards her. Leanne described how, in a row over maintenance payments between Leanne's resident natural mother and her non-resident natural father:

> ... what I did was went into the bathroom and started crying so I didn't make any noise or anything, and then Paul [her step-father] heard me crying. Paul said, "Come out of the bathroom," and so I just came out and he just cuddled me (Interview, April, 1997).

It is interesting here that Leanne's step-father was aware of, and took action to reduce Leanne's distress, whereas, as discussed previously, Leanne's father appeared oblivious to Leanne's disappointments about their relationship. The cuddle indicates her step-father's concern for Leanne, whilst, as discussed previously, the absence of any cuddle from Leanne's non-resident natural father was, to Leanne, symptomatic of the emotional distance between them.

Although Brian was in a children's home until the summer term, 1997, he visited his natural mother and his step-father every weekend. His accounts suggested that he valued his relationship with his step-father. Though Brian, unlike Leanne, Anna and David, did not suggest that he spent time alone with his step-father, his accounts revealed enjoyment of regular family activities, such as trips to the ice-rink and watching videos as a family.

Of all the young people, Martin expressed the most hostility to a resident step-father, as is encapsulated by the uncompromising:

> He's not a dad. He's just a piece of shit (Interview, February, 1997).

It is likely that Martin's unwavering venom towards his step-father was linked to his step-father's physical abuse of him. This is, in turn, similar to the associations between Rebecca's antagonism towards her resident natural father and Louise's and Brian's negativity towards their non-resident natural fathers, and their fathers' violence.

The above therefore suggests that family structure was relatively unimportant in these young people's perceptions of their relationships with fathers, whether natural resident, natural non-resident or step. The descriptions above instead highlight the influences of gender. On the whole, whatever the family structure, these young people presented their relationships with their fathers as lacking the close communication present in some of the young people's accounts of relationships with mothers. Positive relationships with fathers, whether natural resident, natural non-resident or step, tended to centre around shared activities. Some of the young people suggested that their relationships with their fathers, whether natural resident, natural non-resident or step, lacked a sense of connection. As described above, this did not appear to distress either boys or girls in intact nuclear families, although girls appeared more aware of this disengagement than boys. In contrast, the accounts of some girls in reordered families highlighted their distress over their natural non-resident father's apparent disengagement from them. Whereas most of the data suggested the importance of gender and the relative unimportance of family structure in perceptions of relationships with fathers, it would appear that these girls' responses to their natural non-resident fathers were shaped by family structure and gender. In the most starkly negative accounts of relationships with fathers, family structure was unimportant, but there were associations between negative perceptions of relationships with fathers and fathers' violence. In contrast, no young person's account suggested that a mother was violent towards a family member; this further reveals the influences of gender in these young people's accounts of relationships.

SIMILARITIES ACROSS GENDERS AND FAMILY RELATIONSHIPS

The final part of this chapter will focus on case studies of three boys, David, Charley and, Tim. In contrast to the previous discussion of how gender differences in these young people's family relationships might inform academic knowledge about family structure, this section will explore how similarities across genders also potentially inform research on family structure.

David

David (reordered nuclear) has been selected for discussion, because, in his two interviews with me, he focused on his emotions more than most other boys tended to do. In interview, David talked about his feelings about his family break-up, which had taken place two years before fieldwork began. David's distress at the family break-up is captured by the following extract from his first interview:

CH: What are the worst things about families?

David: (long pause) Um, just when um, if you've got, when you 'ave, parents if they like split up.

CH: Why?

David: Because like (long pause) I've been through it, so I I know what it's like, and it's like really horrible, because they're sometimes arguin' all the time and you don't know what to do.

CH: What did you do?

David: Well um started cryin', 'cos I didn't want 'em to argue and I just tried to get to sleep and tried not to listen. There was one time when the neighbour called the police, 'cos 'e

> could hear it as well, an' I was really upset then. I just didn't know what to do, so I ran away (Interview, January, 1997).

In the above, although David could have avoided doing so, he relates my general question on the worst things about families to his own personal situation. David was able to articulate feeling vulnerable and distressed about his family situation. Later in the same interview, David is explicit about having felt unable to handle his family situation:

> It [the family situation] just got worse [with great emphasis] and every night I thought about it and couldn't like, just started cryin', couldn't handle it.

In describing his response to his parents' separation, two years after it had occurred, David demonstrated his awareness of his parents' feelings. Whilst David said that he would have preferred his parents to have stayed together, he also described how both his parents were 'happier' since they had split up, and how, in turn, this had made him 'happy' (Interview, January, 1997).

These extracts from interviews with David are particularly striking because David was usually quiet, both with his peers and, out of interview, with me. Outside his two interviews with me (January and April, 1997), David did not tend to talk at length to me, though he was always pleasant and polite. David's detailed focus, in interview, on his emotions was more similar to girls' accounts of family than to those of most other boys.

Charley

Charley (reordered nuclear) has been selected for discussion because, over a number of interviews, Charley appeared to revel in regaling me with both events in and the interpersonal dynamics of his large immediate and extended family. In contrast to my experience of Charley, teachers at Springfield tended to view him as either silent or monosyllabic; indeed, Karl Price, the Head of Year, found it impossible to believe that Charley could be chatty.

Charley's love of tales about family is perhaps conveyed by his suggestion for a title for my 'book': 'The Family Affairs of Year Nine'. Charley, with little or no prompting from me, would usually develop his narratives about family in some detail, as the following extract from his account of his older half-sister's pregnancy suggests:

> She [his half-sister] was shocked. Mum weren't pleased, but she weren't mad. She wasn't pleased at being a granny at 36. But then she said, "I wanna be the first to see the baby. I'm its grannie. I've got a right." My mum's got this tradition, 'cos my mum was 18 when she had Ruth (Charley's half-sister) and Ruth's 18 now (Interview, April, 1997).

In contrast to most of the other boys, Charley talked openly to me about personal details, as when he told me about his sister discovering she was pregnant after her periods stopped:

> My sister didn't have a period. She went to the doctor's, had a pregnancy test, and found that she was pregnant (Interview, April, 1997).

Charley was also unusual amongst the boys in that he chose to talk in detail about interpersonal dynamics, as in the following, when Charley is describing a crisis in an aunt's and uncle's marriage:

> My uncle was having an affair. Then 'e stopped it and went back to my auntie, but then 'e started seeing 'er behind my auntie's back. My auntie found out, and they 'ad a massive row yesterday, and she chucked 'im out (Interview, April, 1997).

Charley's focus on tales about his family was more similar to girls' accounts of family, than to those of other boys. Furthermore, whilst girls, rather than boys, tended to highlight closeness and caring in interpersonal relationships, Charley was explicit about the importance of these qualities in family life, as the following suggests:

> My mum's boyfriend, the first thing he does after work is see the baby and then ring his dad or else go down his mum and dad's. He cares (Interview, April, 1997).

Charley also showed that he was aware of his mother's feelings, as when Charley described how, at Christmas, 1996, his family waited to open their presents until his mother had come home from hospital, after giving birth to his baby half-sister on Christmas Eve, because:

> . . . she likes to see us openin' them. She likes to look at our faces, and then she knows that we're happy (Interview, February, 1997).

Tim

Tim showed more awareness of his mother's feelings than many other boys did. For instance, Tim was aware of the distress his brother's delinquency had caused his mother, and of how this problem had, according to Tim, exacerbated his mother's thyroid problem. Consequently, Tim declared his intention of treating his mother more thoughtfully than his brother had done:

> My brother's got a long criminal record. My brother steals cars and robs bikes. I'm different, 'cos I've had my brother to show me the example of all the trouble he's in. He made my mum depressed and gave her a thyroid problem. I saw what he'd done to her and I haven't done the same (January, 1997).

It should, however, be stressed that Tim saying he would not upset his mother is not to suggest that, in actuality, Tim never distressed his mother! The interview continued with Tim's accounts of his own activities which, it is not unreasonable to suppose, may have upset his mother:

I steal from shops. That's okay, but not from someone's house, 'cos that's what they've earned. And I smashed lots of windows at [name of a primary school in Castletown] so now I'm not allowed near.

The detailed narratives of family events, emotions and the interpersonal dynamics between different family members discussed above were, on the whole, more characteristic of girls' accounts than of boys'. In this way, David's, Charley's and, to a lesser extent, Tim's accounts contrast with the literature on gender, discussed earlier in this chapter, which highlights differences between genders. David's, Charley's and Tim's cases underline how some claims within the gender literature, such as how women are more empathetic and caring than men and how women focus more than men on interpersonal relationships and on connection with others, are too clear-cut. These three boys, in contrast, affirm Freedman's (1990) emphasis on recognizing similarities as well as differences in considering gender.

The three boys were all from reordered families. Their awareness of their own emotions, of others' feelings and of interpersonal dynamics may be related to Walczak & Burn's (1984) claim, discussed earlier in this chapter, that some research subjects had developed emotional maturity, through the experience of family reordering. These three boys' awareness of self and/or others would appear an attribute, in comparison to much of the research literature's focus on the negative consequences of family reordering. David's, Charley's and Tim's affirmation of Freedman's (1990) emphasis on recognizing similarities across genders illustrates that the literature on gender might contribute to academic knowledge about family structure, by suggesting that some claims about family reordering are, in their turn, too sweepingly negative.

CONCLUSIONS

This chapter has discussed methodological and substantive findings on gender within an ethnographic study of the influences of family structure (intact nuclear, reordered nuclear or step, and single-parent) upon 32 young people's experience of family and schooling, to explore the ways in which consideration of gender might inform academic debate on family structure. As a basis for its argument, the chapter has used two strands within the academic literature on gender: claims of differences according to gender and claims of greater complexities in gender relations than are suggested by a focus purely on difference.

The chapter has shown how ethnography offered opportunities to reduce potential gender imbalances in field relations, so that, as far as possible, both boys and girls were able to speak about family relationships from the basis of relationships of trust with me.

The chapter has shown how these young people's accounts revealed greater closeness and communication in their relationships with mothers than with fathers. This was particularly the case with the girls in the sample. Family structure appeared, in contrast, unimportant in these young people's accounts of relationships with mothers, except that some young people in reordered families appeared particularly aware of their mothers' needs and feelings.

The chapter has argued that there were similarities across the family structures in these young people's accounts of relationships with fathers, whether natural resident, natural non-resident or step. Good relationships tended to be characterized by shared activities with fathers. Some young people's relationships with fathers were characterized by disengagement, in contrast to the young people's relationships with mothers. There was an association with some young people's hostility to fathers and paternal violence. These similarities in perceptions of relationships with fathers across the family structures are particularly important because of the argument in the literature that family break-up impacts negatively upon children's relationships with their fathers. The findings in this ethnography highlight Burghes's (1994) point that more research is needed into young people's relationships with their fathers.

However, the chapter discussed one association between family structure and perceptions of relationships with fathers. Whilst some girls in intact nuclear families appeared more aware of a resident natural father's disengagement than boys, this did not seem to distress them. In contrast, two girls in reordered families appeared distressed by their natural non-resident fathers' lack of involvement with them.

The chapter used case studies of three boys from reordered families to point to similarities across the genders in the young people's accounts of emotions and of family life. It was suggested that these similarities potentially inform academic knowledge on family structure, through the association of family reordering with positive qualities: self-knowledge and awareness of others. This is potentially important because, again, it runs counter to explorations of the predominately negative impact of family break-up in the literature on family structure.

Overall, in this data set, the differences between and the similarities across gender illustrate Thorne's (1990) and Deaux & Major's (1990) point that, in any analysis of gender, it is important to recognize the complexity of gender relations, and Freedman's (1990) argument that any framework for presenting gender needs to be flexible enough to present differences and similarities. In this data set, the existence of both differences between and similarities across genders underlines the influences of gender upon young people's experience of family structure. This contrasts with how much research on family structure has omitted to draw upon the existing body of literature on gender, and points to the potential pitfalls of compartmentalizing academic knowledge. The discussion of gender in this chapter emphasizes the extent to which academic research on gender may inform, rather than being seen as distinct from, academic research on family structure.

REFERENCES

Burghes, L. (1994). *Lone Parenthood and Family Disruption – The Outcomes for Children.* Occasional paper 18. London: Family Policy Studies Centre.

Cockett, M., & Tripp, J. (1994). *The Exeter Family Study.* Exeter: University of Exeter Press.

Deaux, K., & Major, B. (1990). A Social-Psychological Model of Gender. In: D. Rhode (Ed.), *Theoretical Perspectives on Sexual Difference.* New Haven: Yale.

Finch, J. (1984). It's Great to Have Someone to Talk to: Ethics and Politics of Interviewing Women. In: C. Bell & H. Roberts (Eds), *Social Researching: politics, problems, practice.* London: Routledge and Kegan Paul.

Freedman, E. (1990). Theoretical Perspectives on Sexual Difference. In: D. Rhode (Ed.), *Theoretical Perspectives on Sexual Difference.* New Haven: Yale.

Furstenberg, F. F. Jr., & Cherlin, A. J. (1991). *Divided Families: The Family and Public Policy.* Harvard: Harvard University Press.

Gilligan, C. (1982). *In a Different Voice.* Cambridge: Harvard University Press.

Hudson, C. (1999). *Young People's Experience of Family and Schooling: how important is family structure?* Unpublished doctoral thesis, University of Oxford.

Kiernan, K. (1998). Family Change: Issues and Implications. In: M. David (Ed.), *The Fragmenting Family: Does It Matter?* London: IEA.

Maltz, D., & Borkner, R. (1998). A Cultural Approach to Male-Female Miscommunication. In: J. Coates (Ed.), *Language and Gender: A Reader.* Oxford: Blackwell.

McKeganey, N., & Bloor, M. (1991). Spotting the Invisible Man: The Influence of Male Gender on Fieldwork Relations. *British Journal of Sociology, 42*(2), 195–210.

Mitchell, A. (1985). *Children in the Middle. Living Through Divorce* London: Tavistock.

Morrow, V. (1998). *Understanding Families: Children's Perspectives.* London: National Children's Bureau.

Rich, S. (1990). Daughters' Views of their Relationships with their Mothers. In: C. Gilligan et al. (Eds), *Making Connections: The Relational World of Adolescent Girls at Emma Willard School.* London: Harvard University Press.

Rodgers, B., & Pryor, J. (1998). *Divorce and Separation. The Outcomes for Children.* York: Joseph Rowntree Foundation.

Silverman, D. (1993). *Interpreting Qualitative Data Methods for Analysing Talk, Text and Interaction* London: Sage.

Tannen, D. (1998). Talk in the Intimate Relationship: His and Hers. In: J. Coates (Ed.), *Language and Gender: A Reader.* Oxford: Blackwell.

Thorne, B. (1990). Children and Gender: Constructions of Difference. In: D. Rhode (Ed.), *Theoretical Perspectives on Sexual Difference.* New Haven: Yale.

Utting, D. (1995). *Families and Parenthood: Supporting Families, Preventing Breakdown.* York: Joseph Rowntree Foundation.

Walzak, Y., & Burns, S. (1984). *Divorce, the Child's Point of View.* London: Harper and Row.

BEING 'ONE OF THE LADS': INFANT BOYS, MASCULINITIES AND SCHOOLING

Christine Skelton

INTRODUCTION

Research into masculinities and schooling has shown that schools are sites in which a multiplicity of masculinities are played out. For example, the concept of identity which has taken a central part in academic and political discussion and debate in the 1990s (Haywood & Mac an Ghaill, 1997) has prompted investigations into the formation of masculine identities in school settings (Connolly, 1998; Jordan, 1995). These studies have indicated the way in which education influences constructions of masculinities; that is, masculinity is organized on a macro scale around social power, but the education system in this society is such that access to social power, in terms of entry to higher education and professional careers, is available only to those who possess the appropriate 'cultural capital' (Bourdieu, 1986). It has been argued by some commentators that those boys who are unable to obtain entry to the forms of social power schooling has to offer then seek alternative means of publicly demonstrating their masculinity, such as the use of violence or sporting prowess (Back, 1994; Segal, 1990).

There has been a tendency to focus on masculine identities in secondary education, particularly the impact of the ways in which boys are *organized* in relation to the curriculum (e.g. streaming; academic versus vocational subjects. See Connell, 1989; Mac an Ghaill, 1994). These writers argue that schooling is

Genders and Sexualities in Educational Ethnography, Volume 3, pages 91–110.
Copyright © 2000 by Elsevier Science Inc.
All rights of reproduction in any form reserved.
ISBN: 0-7623-0738-2

probably not the key influence in the formation of masculine identities for most males, rather it is sexual relationships and the adult workplace which have a greater significance. However, these issues are in the future for infant boys and, as they are at the beginning of their school careers, they have yet to see themselves as 'school successes' or 'school failures'. Whilst boys start school having already begun the process of constructing and negotiating their masculine identities in the home and amongst friends in the local culture, schooling will necessarily have an impact on the shaping of masculine identities although in different ways at different points in their progression through school.

This chapter will explore the ways in which a group of 6–7 year-old boys in one primary class negotiated masculine identities in the school setting through various discursive positions such as being a *boy, white, child, school pupil,* and a member of the so-called '*underclass*' (Collier, 1995; Morris, 1994; Williams, 1997).

The chapter has five sections. The first section outlines the methodological framework used in the study. The second section provides information about the school, the local environment and the boys themselves. This is followed by a consideration of the 'knowledge' boys brought with them to school regarding being a 'lad'; that is, the culturally exalted form of masculinity predominant in the local area. The fourth section explores some of the discourses within which the boys were positioned and, in particular, examines the tensions between being *a boy* and *a school pupil*. Here, and in section five, the focus will be on a boy called Shane, as the critical incidents generated by his actions and behaviour provided a means of exploring the 'apprentice lad' mode of dominant masculinity as constructed and negotiated by the boys in the class. The final section (five) reflects on where boys positioned themselves with regard to the 'apprentice lad' form of masculinity and in their relationships with each other.

THE CASE STUDY

The data for this paper are taken from a year-long ethnographic study of the construction of masculinities in two primary schools. The school that features in the discussion here is situated in Wickon, an economically deprived area of the north-east city of Oldchester.

My role in the school was that of part-time researcher/part-time teacher. Initially I was disturbed about having to undertake the research wearing the label of 'teacher' as the intention had been to be an 'adult helper' in the classroom. The purpose of situating myself as 'adult helper' was that I wanted to minimize, wherever possible, the number of different power relationships I would have with the boys, girls, teachers and other adults in the school. On reflection this seems rather a naive intention; relationships in the field are established on the basis of not who the researcher 'pretends' to be but rather on the constant construction and negotiation of personal identity. Also, as researchers, we are located and positioned in many different ways; but at the same time we also locate and position ourselves, although this is always defined by one's history, nature, age, gender, 'race', sexuality, social class, etc. (Alldred, 1998; Skeggs, 1994; Stanley, 1997). Thus, to take on a role as adult-helper/researcher in a classroom was unlikely to have prevented me, for example, from drawing on the understandings I have about teaching and learning or managing and organizing children in the classroom.

A further matter was the extent to which I, as an adult woman, could access information about the ways in which young boys construct and negotiate their masculine subjectivities. Subjectivity is constructed across a range of sites and even if I could have accessed the kinds of information about how masculinity is constructed in relation to other masculinities (as men researchers have been able to; see; Connell, 1989; Connolly, 1998; Mac an Ghaill, 1994; Walker, 1988) this would still not have provided information about boys' positioning within other discourses such as son, friend, child, etc. It was the issue of accessing masculine subjectivities, together with my feminist 'sensitivities' and the fact that the schools had determined my role as that of part-time teacher/researcher, that established the focus of the research as one which used the *schools'* constructions of modes of masculinities in order to explore how boys engaged with, challenged, negotiated, rejected and re-constructed their masculine selves in school settings. Thus my research questions were, "What part does the primary school play in constructing, challenging and re-constructing forms of masculinities and male practices?" and, "If schools are sites where multiple modes of masculinities are constructed, negotiated, challenged and re-constructed what do these 'look like' in terms of male actions, behaviours and attitudes in the primary classroom?"

At the centre of this research was the issue of male power. The political will underpinning the research was to gain insights into how boys engage with male power in the forms it takes in primary schools in order to add to feminist understandings. Although criticized as problematic, the concept of 'hegemonic masculinity' proved useful in that it allowed for the unbalanced nature of gendered power relations in schools to be explored whilst recognizing that the dominant position is one that has to be constantly won (for criticisms see Kerfoot & Whitehead, 1998; Macinnes, 1998). Crucially, although hegemonic masculinity is about power, the most visible bearers of hegemonic masculinity are not always the most powerful people (Connell, 1995). As such, a theoretical framework was required which could facilitate an understanding of hegemonic masculinity at a number of levels. This, of course, led to the dilemma of how to link the findings of a small scale study to larger-scale theories – the issue which is central to the micro – macro debate.

A number of writers have argued for a theoretical framework which makes use of a range of positions (Haywood & Mac an Ghaill, 1997; Troyna & Hatcher, 1992; Weiner, 1994). A model which retains the significance of ideological structures in the construction of identities was developed by Troyna and Hatcher (1992) for their work on racism and primary school children. This model attempts to avoid the problems of reading off macro explanations from small-scale studies by looking instead at how various societal aspects interrelate and inform specific instances. The levels of this model are: political/ ideological; cultural; institutional; sub-cultural; biographical; contextual; interactional. It is important to note that the levels link with each other but there is no implication of chronological development or linear flow (see Fig. 1).

Before moving on to the findings of the case study a number of points need to be reiterated or acknowledged. First, only a partial picture of boys' masculine identities at school can be offered. The findings presented here provide a partial picture, in that the focus in the study was on the dominant mode of masculinity given ascendancy in the *school* itself. Thus, there is no claim to having made insights into how individual boys at the school made sense of their masculine subjectivities. Second, it is beyond the scope of this paper to engage in debates concerning the concept of hegemonic masculinity but in order to recognize that such discussions are taking place then the phrase 'dominant masculinities' is used as an alternative where appropriate. Third, throughout this paper the words 'masculinity' and 'masculinities' have been used interchangeably. When either is used it can be taken that the plurality of 'being, knowing, understanding and enacting' maleness in relation to structural, collective and individual male practices is understood and recognized.

THE SCHOOL AND CLASS B BOYS

As mentioned above, the school is situated in Wickon, an economically deprived area of Oldchester, in the north-east of England. The area has witnessed the closing of traditional industries and, as a result, experienced steep increases in male unemployment. Alternative forms of work have emerged and the 'cultural economy' is one in which activities from petty thieving to more organized crime have flourished (Campbell, 1993). The

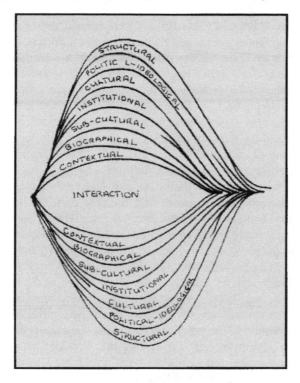

Fig. 1. A model for analyzing hegemonic masculinities in schools.
Source: adapted from Troyna & Hatcher (1992, p. 40).

people committing these crimes are predominantly young men in the 14–19 age bracket (Phillips, 1993). At the time the fieldwork was undertaken there was a heavy police presence in the area. The media too were constant visitors as Wickon frequently appeared in television or Sunday newspaper articles as representative of those areas which house the 'underclass'.

The school was a target for theft and vandalism and a number of measures were adopted for securing the premises, such as heavy metal gates and doors, security cameras and several entrances being locked during school hours. There were approximately 370 children on the roll, in 13 classes and a nursery. The research focused on the 24 children in Class B, a Year 2 group (aged 6–7). The class was composed of 10 girls and 14 boys, all white, and with one exception, a girl from Scotland, all born in the north-east. The class teacher was Terry Blake and the headteacher was Mrs Masterson.

There were four boys in Class B who were immediately noticeable: John, Shane, Luke and Robert. John and Shane made their presence known to any newcomer, adult or child, to the class. They took the lead in encouraging behaviours amongst the other boys designed to 'suss out' (Beynon, 1984) unfamiliar teachers; for example, responding to a query or instruction by ignoring or answering back, and engaging in general 'tomfoolery' rather than work. During the course of the year these two boys were, at various times, the best of friends and the worst of enemies. The tensions between them occurred largely as a result of their intense rivalry for what they referred to as the 'first boss' position. Both boys came from families who had a high profile in the Wickon community in terms of their notoriety. During the time of the research both families had substantial contact with the police. At the beginning of the observation period one of Shane's brothers was in a remand home. He escaped from the remand centre on three separate occasions, and each time was re-arrested at the house of one or other of his aunts. John's father was also in court, involved in a protracted trial; this information was given to the school by John's mother, although no indication was given regarding the crime.

Luke was keen to associate himself with the activities of John and Shane but was hampered in his attempts to 'lead' as he seemed universally disliked by boys, girls and teachers. The other children listed the reasons for disliking him as 'fat', 'smelly', 'snotty-nosed' and "having skid marks on his pants." Carl and Rick were also marginalized within the boys' group because of their physical appearance and personal habits; Rick because he was 'smelly' and Carl because he was 'slavery' (he dribbled). As a result of being avoided by many of the children in the class they tended to seek each other out at 'choosing time'.

Robert stood out in a totally different way. He had recently moved into the Wickon area from a part of Oldchester, also economically deprived, but which did not have the same reputation for crime. In contrast to all the other boys, he seldom indulged in disruptive behaviour and certainly never instigated it. Robert was described by both Terry Blake and Mrs Smith, the other Year 2 teacher, in terms of being a 'really nice boy'.

The remaining eight boys can be loosely grouped in terms of their relationship with John and Shane. As was said earlier, it tended to be those two boys who instigated the majority of competitive, challenging behaviours both to authority and amongst their peer group. Gary, Tommy and Matt were always the first to join in with John and Shane and, occasionally, attempted to initiate and take the lead in various challenging actions. Bobby, Adam, Dean, Sean and Martin always took part in any group actions but were unlikely to lead. Observing the way in which the boys tended to take up the same positions in the group whenever a challenge was mounted to a teacher's authority or for a confrontation with boys from another class, reminded me of the organization of the army during the Second World War.

Although a very crude analogy, it provided a useful image of the boys' metaphorical and often literal positions in the group:

- the generals (John and Shane), who organized the action and led the initial assault;
- the regular soldiers (Luke, Gary, Tommy and Matt), who were quick to see what was required and proficient in supporting the actions of the leaders;
- the conscripts (Bobby, Adam, Dean, Sean and Martin), who realized they had to join in but their involvement was minimal and they literally positioned themselves on the periphery of the action;
- the group who could have been seen as conscientious objectors (Robert, Rick and Carl), although for different reasons. Robert's preferred style of challenge was verbal whilst Rick and Carl recognizing their marginal position amongst the boys preferred to avoid contact with the rest of the group. At the same time Robert, Rick and Carl (and indeed the 'conscripts') always took some role in any action, possibly because the alternative was more personally threatening.

The boys could be discussed as a 'group', on the basis of a shared relationship to the dominant masculinity of the local culture/school where they constructed and negotiated masculine identities at various levels. Although the boys in Class B were too young to access the forms of power utilized by the 'lads' in the local community, they were aware of the power and status the 'lads' held and, as males themselves, knew they would eventually gain entry to this fraternity. Indeed, as will be shown, the actions of the boys, particularly those of Shane and John, can be interpreted as ways of 'working' themselves into older forms of masculinity in preparation for this time (Redman, 1996).

BOYS, MASCULINITIES AND THE LOCAL CULTURE

The 'hard man' image of masculinity occupied a high status in the Wickon area, particularly amongst the 'lads': gangs of 14–16 year olds whose actions and presence on the streets in the local community attracted constant police attention (Campbell, 1993; Phillips, 1993). The intention here is to consider the implications of being a 'hard man' for the interrelationships of the 'lads'. The purpose of this is to provide insights into the particular dominant masculinity of the 'lads' which the boys in Class B would have become familiar with; a familiarity which would help shape their own masculine identity.

Being a 'Lad'

The 'lads' hung around together, but the evidence from studies of the area indicates that there was always a leader or leaders who managed the alternative, unconventional forms of 'work' which they engaged in. The 'cultural economy' was one in which petty thieving to more organized crime held a central place. A mother living in the area observed:

> There is a hard core of 14–19-year olds and usually an older man, about 25 or 30, who encourages them. It's dead hard keeping your kids away from street culture. These boys belong with each other. They bond to each other. They brag about how they get away with what they do . . . They hear, so the story goes, that they can make up to £300 a week with petty thieving and far more through organized crime. These kids see that as a legitimate goal (Quoted in Phillips, 1993, p. 33).

One of the ways to achieve high status within the group was via the confrontations with authority structures, notably the police. Particular kudos could be gained if their activities attracted media coverage. So, local and national television footage of high speed car chases and, on one occasion, the theft of a well-publicized 'thief-proof' police car from the police station car park, brought credibility and status amongst peers and 'admiration and respect from younger ones' (Wallace, 1992, p. 28). The intra-group struggle for a position in the hierarchy has to be seen in relation to inter-group, cultural/ regional contestations.

Being a 'lad' also had racist overtones, For those males in Wickon who were anything other than white meant being positioned as a form of *marginalized* masculinity (Connell, 1995).

Given the large family networks living in Wickon, it is likely that the boys in the sample were familiar with what was entailed in being a 'lad'. The characteristics that allowed access to 'laddish' culture were being tough, competitive and white. Being a 'lad' also involved taking on a certain role as a member of a gang or group. The boys in the class had probably observed their male relatives engaged in struggles over hierarchical positionings in their various gangs or groups. Indeed, the fact that the males in the families of both John and Shane occupied a 'ringleader' status was reflected in their own struggles to be head of the boys' group.

Similar to Connolly's (1998) 'Bad Boys' whose masculine identities were constructed and negotiated in a culture concerned with day-to-day survival, so too the boys in Class B were engaged in constructing masculinities informed by discourses of boyhood and childhood particular to the local context.

Discourses on Being a Boy and Childhood

In discussing discourses on *boys* and *childhood* prevalent in the local culture, the intention is not to create an artificial divide, as clearly there is a dialectic. Research literature has shown that young children invest a great deal of effort in establishing themselves, personally and publicly, as a member of their gender group (Davies, 1989; Jordan, 1995). Being a male in Wickon was associated with forms of power (albeit subversive/anti-authority forms). So whilst young boys were not in a position to access those avenues of power, it was something which would be available to them in the future.

On the other hand, being a young child in the Wickon area appeared to be a necessary but undesirable phase; that is, where babies were proudly paraded round the streets in their prams and adolescence heralded new-found status, childhood was framed in a similar way to that of working-class children in Victorian times and earlier, where the emphasis was on practising to be an adult (Cunningham, 1991). For example, the children in Class B sported hairstyles, clothes and occasionally, with the girls, makeup, which reflected the fashions of adolescents. Children might also be encouraged to drink alcohol and smoke. Shane informed the class one 'news time' that he had been to the local pub where his dad and his mates had given him some beer. Whilst there have to be some reservations regarding the accuracy of such comments, given the boys' desires towards 'laddishness', some observations suggested these were not necessarily fabricated accounts. One morning Adam came into school looking pale and saying he felt sick. He said his uncle had put a 'tab' in his mouth and told him to suck it. Adam had done this but it 'tasted horrible so I chucked it on the floor and stood on it. Me uncle was cross 'cause it was a new 'un' (*Field notes*).

These factors contributed to shaping the boys' knowledge, awareness and construction of their own masculine identities, a process which continued in the more public arena of the school. Here the boys were confronted with ambiguities identifiable between the school and local cultural discourses on childhood. Furthermore, they also encountered the tensions involved in discourses on being *a boy* and *a school pupil*.

PRIMARY SCHOOLING AND BOYS

The 'Nature' of Childhood

Beliefs and assertions about childhood and the nature of children have been, and continue to be, important elements in the professional ideology of primary school teachers (Pollard, 1987). One line of argument suggests that these beliefs contain ambiguous conceptions of children; that is, between teachers' societal and individualistic aims (Ashton et al., 1975). From this perspective, teachers' societal aims position children as immature, irresponsible and dependent and, therefore, pupils need to be taught certain things for the benefit of society. On the other hand, teachers' individualistic aims stress the importance of personal growth and self-expression for their pupils. Whilst this is clearly only one view of teachers' professional ideology, there was evidence of some ambiguity when the teachers discussed their aspirations for their pupils. Teachers linked the development of self-esteem (individualistic aims) with the ability of pupils to take responsibility for themselves and their actions (societal aims):

> Terry Blake: . . . one of my goals for next year will be to try and raise their self-esteem more . . . At the beginning of the year when I gave them a task there were a good eight or nine children who would just sit at the table and do nothing until I actually spoke to them individually . . . that happens less and less . . . I encourage them to work and think for themselves . . . I suppose that's perhaps the aim, to get them to think more for themselves about what they're doing . . . to take responsibility for their learning.
>
> Mrs Masterson: My aims are for the children to come to some sort of ableness . . . to think for themselves, to engage in their own learning. That means raising their self esteem, their self confidence as the parents don't. I'm not bothered about the National Curriculum . . . they'll get that . . . it's *how* they learn, getting them to think, getting them to take responsibility for themselves, that's important.

The stress given here to developing a sense of 'responsibility' in the children seemed to have resonances with parental aims. Encouraging 'responsibility' could be interpreted as the school's desire for pupils to develop independence and show maturity. Similarly, those parents who dressed their children as adolescents and encouraged adult behaviour such as smoking and drinking might also be seen to be promoting maturity/independence. However, as implied so far, the reasons for, and means by which, this independence and maturity was fostered, differed between teachers and parents. In fact, the strategy teachers employed to encourage boys to take responsibility had an unintentional consequence, as will be shown in the section on *Self-esteem and Masculinities*.

As can be seen from the quotations above, teachers linked the development of pupils' self-esteem with a greater sense of responsibility. Not only was self-esteem believed to facilitate pupils' ability to take responsibility for their learning, but much of the psychological literature on classroom behaviour links negative behaviour with low self-esteem (Fontana, 1988; McIntire, 1984). Therefore, it could be argued that if pupils acquired higher self-esteem then a consequence would be increased conformity to classroom rules. However, the boys in Class B were positioned by multiple discourses and a particularly powerful discourse was of being a 'lad' or, in their case, an 'apprentice lad'. The tension created was that being a *school pupil* and being a *lad* demanded conflicting behaviour. Whilst developing self-esteem might encourage conformity to classroom rules in the boys as *school pupils,* the opposite (lack of conformity) was expected of the boys as 'apprentice lads'.

School Boys and 'Being a Boy'

For the teachers, taking responsibility meant pupils recognizing and conforming to the school authority structures; but for the boys *as* boys, 'responsibility' had implications for power dynamics, as the work of Alloway & Gilbert (1997) testifies. The question here is how the tensions between discourses on *school pupil* and those in the local culture of *being a boy* were manifested. A useful starting point is to consider the boys' understandings of what was important to them, as boys, in the classroom. The following conversation suggested their main concern related to their position in the masculine hierarchy:

Luke: Who's the boss?

Shane: I'm the boss!

John: *I'm* the boss!

Luke: One can be first boss and one can be second boss.

Shane: Who's first boss and who's second boss?

Robert: You can take it in turns.

Shane and John (in unison): I'll be first boss!

Robert: Shane is first boss and John can be second boss.

Luke: Yes.

Shane: I'll be first boss, John'll be second boss . . .

Luke: I'll be third boss and Robert can be four boss . . .

John: Shane'll be first boss, I'm second boss, Luke three boss and Robert four boss.

Robert: I don't want to be four boss.

Whether Robert's protests were because he did not want to be *fourth* boss, or because he did not want to be considered a *boss* at all, is not known, but either way it is not pertinent to the issue raised here. It can be argued that two different agendas were in place. For the teachers, the aim was to foster a sense of responsibility both academically and behaviourally in the boys (and girls) as *school pupils,* whilst for the boys the main concern was to establish their masculine identity and place in the male hierarchy. The next section will explore this by focusing on how the ways in which teachers' attempts to develop self-esteem in Shane inadvertently contributed towards his struggle for 'first boss' position.

Self-esteem and Masculinities

Mac an Ghaill (1988, 1994) has shown how the 'Macho Lads' in his study linked teacher and police authoritarianism and, as a consequence, developed their particular 'tough' version of masculinity around collective strategies of counter-interrogation, contestation and survival. Although it is fair to say that the boys in the class were probably aware from an early age of the similarities in the roles of the police and teachers in terms of control and discipline, the latter did occupy a different position. Terry Blake and Mrs Masterson referred in their interviews to the ways in which mothers would use the school as a threat by saying to pre-schoolers they would not be able to act in 'that' way when they went to school, and to older children, that they would 'tell the teacher' about the child's behaviour. It has been argued elsewhere that relationships between the teachers in the school and the local community were partly based on notions of maternalism/paternalism and partly by fear/ vulnerability (Skelton, 1996) and a similarly conflictual position could be observed in relationships between the boys and teachers. As will be shown in the section on Inter-Group Conflict, the young age of the boys meant that they would look to adults (teachers) as a source of comfort and protection. Yet in the same way that the 'lads' would stage events or engage in activities that would enable them to demonstrate superior skills, such as driving faster than the police in stolen cars, the boys in the class would 'try out' the authority of the teacher.

Shane and John did not aspire to secure the approval of their teachers for conformity and 'good' behaviour, but to gain the recognition of the other boys and adults in the school of their potential as a 'lad'.

As the year progressed it became evident to teachers that it was Shane who would consistently attempt to bend, rather than always break, classroom rules, whilst John often seemed simply too tired to bother. To address this teachers attempted to work on Shane's self-esteem by involving him more centrally in classroom life with the result that he was constantly being singled out for attention:

Terry Blake is reading a story to the children who are sitting on the carpet. Adam, Vicki and Katy have each asked over the last few minutes if they can go to the toilet (the rule is only three children are allowed in the toilets at any time). John stands up and walks past Terry Blake saying he is going to the toilet. Terry Blake calls him back and says he must wait. John sits back down. After about ten seconds he stands up and goes to the toilets even though none of the other three has returned. He stays out of the room for about ten minutes but continues to pop his head round the door to smile at the others (Terry Blake has his back to the door) (*Field notes*).

Mrs Cooper (part-time teacher) reads *The Enormous Crocodile* to the class. The children have enraptured expressions on their faces. She shows the illustrations to them and asks "Which one is the crocodile? Shane, can you show us?" Shane smiles, stands up, comes across the carpet, points to the crocodile then roars, pretends to be the crocodile and runs up to other children to 'eat' them. When he eventually calms down, he sits down. No other child is asked to contribute throughout the story (*Field notes*).

These strategies may well have been building up Shane's self-esteem; the assumption, however, that high self-esteem is inextricably linked with the conformist behaviour that schools value is ignoring wider contexts. As Dorothy Rowe (1994) has argued:

It's absolutely true that to survive you have to have something you think you're good at. But you can see this in schools where kids who don't achieve find they *are* good at getting away with things and not getting caught. To say that's not self-esteem as you define it, is just imposing white, middle-class values (Rowe, quoted in Grant, 1994, p. 23).

The real achievement for Shane was in being able to demonstrate to his peers that not only could he take on and outwit the teachers, but that they actively appeared to sanction his behaviour by giving him more attention. Even being caught in attempts to outmanoeuvre teachers added to his self-esteem because it was important, as a 'lad' to be seen by his mates as having attracted attention to himself. As Campbell (1993) has argued, being caught and prosecuted, particularly if it involved a court appearance, was not seen as failure but added to the individual lad's status amongst the others.

The constant challenging behaviour of Shane towards teacher authority earned him a significant amount of teacher attention, and thereby the attention of the other boys. Attempts by teachers to encourage conformity through developing his self-esteem placed him even more centrally in the spotlight. Indeed, Shane's aim of being 'first boss' appeared to be enabled by these actions which pointed to the contradiction between the intended and actual aims of the teachers as a source of critical incidents. As will be shown in the next section Shane was able to utilize teachers' attempts to promote his self-esteem in securing his 'first boss' place amongst the other boys.

BOYS' RELATIONSHIPS WITH EACH OTHER

A Boy Amongst Boys

Dominant modes of masculinity do not reside within individual personae but are 'ideal' constructions which few can actually achieve. The specific form of dominant masculinity in the local culture revolved around being a 'real' (hard) man, and this inevitably demanded rigorous, exacting standards of the boys. The boys in Class B were engaged in running the risks involved with being in, what Connell (1995) has referred to as, the frontline troops of patriarchy (p. 79). A note of caution needs to be sounded here to avoid conflating the terminology Connell uses to define hegemonic masculinity with the particular situation in which it was manifested in the local culture. That is, it is not appropriate to 'read off' the idea that dominant (violent) modes of masculinity are specific to working-class masculinities. Such a perception fails to take into account the fine-grained and complex ways in which masculinities are constructed. Indeed, as was evident in this research, the fact that some of the boys organized themselves in *relation* to the particularly dominant mode of masculinity practiced by John and Shane suggested that alternative forms of masculinities were operating around and within it.

Taking into account the above proviso, it can be argued that as the 'hard man' mode of dominant masculinity centred around violence, aggression and competitiveness, then struggles to construct and negotiate one's individual male identity within this frame involved constant confrontations and challenges *between* men/boys. This positioning in a masculine hierarchy can be shown through an exploration of the interpersonal relationships of the boys in Class B.

Shane's success in outmanoeuvring the authority of the teacher was made evident when others would join in with a situation which he initiated. The boys in Class B did not constitute a gang as defined in the literature (Goldstein, 1994; Miller, 1982; Short & Strodtbeck; 1965). Also, there is no one set of characteristics associated with being a gang leader (Patrick, 1973), although one esteemed study has argued that "The leader is usually . . . the best organizer and planner of delinquent activities" (Haskell & Yablonsky, 1974, p. 174). One such incident involved a class mathematics lesson where the teacher described a shape and the children had to decide which one it was. Shane was asked to begin but then interjected when it was other people's turns. After the third interjection he was rebuked by Terry Blake but continued in his attempts. When Gary, followed by John and Matt, attempted to mimic Shane's success at getting 'one over' the teacher they were immediately curtailed. Shane then changed his strategy:

> The teacher has turned to deal with a message that has been brought in. Shane does press-ups on the floor behind the teacher's back. Martin joins in by demonstrating karate kicks but Terry Blake sees him and tells him to sit down (*Field notes*).

Some studies have found a hierarchical structure to the gangs in some communities with boys progressing from 'toddler' gangs through to 'heavy' teams (Kobrin, 1962; Patrick, 1973). Certainly the behaviour of Shane and John in seeking 'first boss' position, and the willingness of the other boys to participate in subversive classroom activities, suggested their actions were in keeping with those of the 'lads'. As such, competing with each other to show who had the abilities to outmanoeuvre the authority of the teacher was not the sole means through which the boys negotiated their place in the hierarchy or their masculine relationships with each other.

The Role of Humour

Whilst the 'lads' went to great lengths to demonstrate their individual skills at undermining police authority, many of their activities were generated as the result of collective practices, so establishing a shared basis was equally important; that is, the 'lads' would 'hang out' and have a laugh together (Campbell, 1993). A significant feature of the studies of working-class masculinities is the importance of humour to 'macho' forms of masculinity (Corrigan, 1979; Mac an Ghaill, 1994; Sewell, 1997; Willis, 1977). The favoured verbal game-play was 'telling the teacher'. This was where a boy would attempt to get another into trouble by 'telling':

> Dean and Bobby are sitting next to each other during a wet lunch-time watching a *Ninja Turtles* video. They are fun fighting. Dean starts to grab at Bobby's legs and tickling his neck. Bobby pushes him and shouts out "Miss, Miss, he's hitting me!" The teacher responds "Oh Dean! Stop that now!" Both boys laugh and turn their attention back to the video (*Field notes*).

'Telling the teacher' as a form of humour and 'fun fighting' offered a means through which the boys could establish bonds with each other, and this enabled the kind of collective practices towards authority discussed earlier. At the same time, the boys did not always get on together and occasionally, unlike older boys in the school, the teacher's authority would be sought.

Inter-Group Conflict

Although Shane was undoubtedly the most enthusiastic 'apprentice lad' he was only six years old. His struggle to construct a masculine identity as 'first boss' sometimes appeared to overwhelm him and he would call on Terry Blake for support. Observations of what happened between boys after the class teacher had been asked to intervene suggested that the boy whose behaviour had been complained about always made some retaliatory gesture, as in the following episode:

> John takes one of Gary's words he is using to make a sentence. Gary crosses the room and tells Terry Blake. Terry turns and shouts to John that he must give Gary his word back, which he does. Gary sits back at the desk. John watches Terry Blake and, when he is involved with another child, snatches Gary's pencil and throws it on the floor (*Field notes*).

These attempts to have the final say were common, and generally the boys appeared not to harbour resentments. However, with Shane retribution was far more protracted and vicious. Research into gang leaders has identified different characteristics including physical and verbal aggression of the leader (Patrick, 1973). The 'hard man' form of dominant masculinity in the Wickon community was inscribed with violence and aggression and, hence, it was more likely that leaders would draw on these to assert and retain their position. Shane, in his securing of 'first boss' position, not only used similar forms of violent behaviour but, importantly, kept up a sustained attack which drew in other boys in the group. One lunchtime Dean apparently hit Shane with a stick. Shane reported this incident to Terry Blake who said that they were not being very nice to each other and told them to sit down. Shortly after, Shane announced loudly that Dean had his sweatshirt on back to front which clearly embarrassed him. Shane was then observed by Terry Blake hitting Dean in the face and when asked why replied his hand had 'slipped':

> (Later again). Dean is standing near the box of bricks. Shane comes up behind him and pushes him roughly to the floor. Terry Blake turns at the noise and, before he can say anything, Shane says, "I'm only playing." Terry Blake asks "Does Dean know you're playing?" In response Dean stands up smiling which implies that what Shane has said is true (*Field notes*).

> The children have been told to tidy away. Dean is kneeling on the carpet playing with the cars. Shane walks past him, puts his hand on the back of his head and forces his face down into the carpet. He quickly walks away. Dean looks up, rubs his chin and stares at Shane but says nothing (*Field notes*).

Having set up a situation in which he punished Dean persistently over the course of an afternoon, Shane built upon this over the next few days by isolating Dean from other boys who seemed to be offering any form of friendship to him. For example, Bobby was helping Dean with his maths and Shane said, "Man, man come over here . . . this is real hard work, not easy-peasy stuff like that." As suggested here, the main strategy used by Shane was that of ridicule. By ridiculing Dean, as in the comment about the sweatshirt, Shane implied that anyone who hung around with him was also ridiculous. Throughout all this Dean made no attempt to defend himself. Within days other boys had joined in the attacks, justifying their actions on the basis of Shane's alleged mistreatment by Dean. For example, a few days after that afternoon Terry Blake told Dean off for copying his maths work into his weather diary. Adam told me that he, Gary and Martin had told Dean to do it:

CS: Why did you do that?

Adam: Because we wanted him to get wronged by the teacher.

CS: Why? What had he done?

Adam: (shrugs) We think he hit someone. Shane it was.

This section has attempted to outline the ways in which the boys negotiated their masculine identity within the classroom culture. Similar to the 'lads', it appeared they needed to demonstrate both individuality as well as group cohesion, and their effectiveness at this secured a place in the male hierarchy, albeit at various levels in relation to the 'boss'.

CONCLUSION

This chapter has been concerned with the ways in which masculine identities were constructed, negotiated and re-constructed by the infant boys in the school setting. The knowledge and awareness the boys brought into school with them of the dominant masculinity in the local culture appeared to inform their own behaviours and relationships. Where the police provided a focus for anti-authority activities for the 'lads', an awareness of the control/discipline aspect of the teaching role partly informed the boys' relationships with their teachers. In fact, shortly after the end of the observation period Shane and John had made their first attempt at 'getting one over' the police. Terry Blake said that to his knowledge what had happened was that John and Shane had been caught by the police after being seen removing a steering wheel from a car. According to an older boy, they had tried to drive it but neither of them could get their feet on the pedals! Further attempts with one steering and one operating the pedals had also failed, so, rather than go away empty-handed (as reported by Shane at 'news time'), they had taken the steering wheel off, even though they knew they had been seen by a police patrol.

It has also been shown that the process of normalizing masculinity takes place around and within a framework of discourses which the boys drew from and were located within. However, the 'grid of possibilities' (Skeggs, 1991) offered by this framework was itself constructed through power/knowledge positions (Heath, 1982). The available discourses had differing relationships to power, so the discourse on *being a boy* drew on and incorporated greater access to power than discourses on *childhood*, or being a *school pupil;* therefore, discourses on the former were much more powerful (influential). The emphasis has been on those boys who were part of Connell's (1995) 'frontline troops', and it needs to be reiterated that there were boys in the class like Bobby, Dean, Sean, Adam, Martin and Robert who appeared to join in with the activities of the more forceful, challenging boys because it could have been potentially more personally damaging *not* to have colluded. Indeed, in much of the autobiographical literature written by men on masculinities writers recall the fear they felt at school of being accused of being a 'poofter', 'wimp', or 'a girl' (Cohen, 1990; Jackson, 1990; White, 1989). At the same time it is important to emphasize two important points: that violent modes of dominant masculinities are not the 'preserve' of working-class male practices and indeed the responses of boys like Bobby, Sean, etc. indicate that alternative, if not resistant, patterns of masculinities are often operating within a more explicitly evident hegemonic framework. Also, the situation at the school in terms of the articulation of these differing modes of masculinities were particular to this school site and cannot be generalized to other primary schools located in similar economically situated areas.

A problem here is that, in saying the above, this case study can be criticized in a similar way to much of the literature on masculinities and schooling. Whilst the literature on boys' underachievement generates 'quick fix solutions', that on studies of masculinity has been rather hesitant when it comes to offering practical strategies to schools and teachers. Epstein et al. (1998) argue that the reason for this is that "the issues are multi-faceted, the research complex, and it would be premature to suggest firm directions for others to follow" (p. 14). This chapter has sought to support this thesis and, in so doing, implies that individual schools need to recognize the modes of masculinity operating within the school in order to develop responses and strategies which are relevant to their particular situation and needs.

REFERENCES

Alldred, P. (1998). Ethnography and Discourse Analysis: Dilemmas in Representing the Voices of Children. In: J. Ribbens & R. Edwards (Eds), *Feminist Dilemmas in Qualitative Research.* London: Sage.

Alloway, N., & Gilbert, P. (1997). Boys and Literacy: Lessons from Australia. *Gender and Education, 9*(1), 49–59.

Ashton, P. M., Kneen, P., Davies, F., & Holley, B. J. (1975). *The Aims of Primary Education.* London: Macmillan.

Back, L. (1994). The 'White Negro' Revisited: Race and Masculinities in South London. In: A. Cornwall & N. Lindisfarne (Ed.), *Dislocating Masculinity.* London: Routledge.

Beynon, J. (1984). 'Sussing out' Teachers: Pupils as Data Gatherers. In: M. Hammersley & P. Woods (Eds), *Life in School: The Sociology of Pupil Culture.* Milton Keynes: Open University Press.

Bleach, K. (1998). *Raising Boys' Achievement in Schools.* Stoke on Trent: Trentham.

Bourdieu, P. (1986). *Distinction: a Social Critique of the Judgement of Taste.* London: Routledge.

Campbell, B. (1993). *Goliath.* London: Methuen.

Cohen, D. (1990). *Being a Man.* London: Routledge.

Collier, R. (1995). *Masculinity, Law and the Family.* London: Routledge.

Connell, R. W. (1989). Cool guys, swots and wimps: the interplay of masculinity and education, *Oxford Review of Education, 15*(3), 291–303.

Connell, R. W. (1995). *Masculinities.* Cambridge, Polity Press.

Connolly, P. (1998). *Racism, Gender Identities and Young Children.* London, Routledge.

Corrigan, P. (1979). *Schooling the Smash Street Kids.* London: Macmillan.

Cunningham, H. (1991). *The Children of the Poor.* Blackwell: Oxford.

Davies, B. (1989). *Frogs and Snails and Feminist Tales.* London: Allen and Unwin.

Epstein, D., Elwood, J., Hey, V., & Maw, J. (Eds), *Failing Boys? Issues in Gender and Achievement.* Buckingham: Open University Press.

Fontana, D. (1988). *Psychology for Teachers.* Basingstoke: Macmillan.

Goldstein, A. (1994). Delinquent Gangs. In: J. Archer (Ed.), *Male Violence.* London: Routledge.

Grant, L. (1994). Positive Thinking Doesn't Work. *Independent on Sunday,* (May), 23.

Haskell, M., & Yablonsky, L. (1974). *Juvenile Delinquency.* Chicago: Rand McNally.

Haywood, C., & Mac an Ghaill, M. (1997). Materialism and Deconstructivism: Education and the Epistemology of Identity. *Cambridge Journal of Education, 27*(2), 261–272.

Heath, S. (1982). *The Sexual Fix.* London: Macmillan.

Heward, C. (1991). *Public School Masculinities – An Essay in Gender and Power.* Private Correspondence.

Jackson, D. (1990). *Unmasking Masculinity.* London: Unwin Hyman.

Jordan, E. (1995). Fighting Boys and Fantasy Play: The Construction of Masculinity in the Early Years of School. *Gender and Education, 7*(1), 69–86.

Kerfoot, D., & Whitehead, S. (1998). Whither Hegemonic Masculinity? Paper presented to Gendering the Millennium Conference. University of Dundee, 11–13 September.

Kobrin, S. (1962). The Impact of Cultural Factors on Selected Problems of Adolescent Development in the Middle and Lower Class. *American Journal of Ortho-psychology, 32*(3), 387–390.

Mac an Ghaill, M. (1994). *The Making of Men: Masculinities, Sexualities and Schooling.* Buckingham: Open University Press.

Macinnes, J. (1998). *The End of Masculinity.* Buckingham: Open University Press.

McIntire, R. W. (1984). How Children Learn, In: D. Fontana (Ed.), *The Education of the Young Child.* Oxford: Blackwell.

Miller, W. B. (1982). *Crimes by Youth Gangs and Groups in the United States.* Washington: National Institute of Juvenile Justice and Delinquency Prevention.

Morris, L. (1994). *Dangerous Classes.* London: Routledge.

Patrick, J. (1973). *A Glasgow Gang Observed.* London: Methuen.

Phillips, A. (1993). *The Trouble With Boys*. London: Pandora.

Pollard, A. (1987). *Children and their Primary Schools*. Lewes: Falmer.

Redman, P. (1996). Curtis Loves Ranjit: Heterosexual Masculinities, Schooling and Pupils' Sexual Cultures. *Educational Review, 48*(2), 175–182.

Salisbury, J., & Jackson, D. (1996). *Challenging Macho Values*. London: Falmer.

Segal, L. (1990). *Slow Motion: Changing Masculinities, Changing Men*. London: Virago.

Sewell, T, (1997). *Black Masculinities and Schooling*. Stoke-on-Trent: Trentham.

Short, J., & Strodtbeck, F. L. (1965). *Group Process and Gang Delinquency*. Chicago: Chicago University Press.

Skeggs, B. (1991). Postmodernism: What is All the Fuss About? *British Journal of Sociology of Education, 12*(2), 255–267.

Skeggs, B. (1994). Situating the Production of Feminist Ethnography. In: M. Maynard & J. Purvis (Eds), *Researching Women's Lives from a Feminist Perspective*. London: Taylor and Francis.

Skelton, C. (1996). Learning to be 'Tough': the Fostering of Maleness in one Primary School. *Gender and Education, 8*(2), 185–197.

Stanley, L. (1997). Methodology Matters! In: V. Robinson & D. Richardson (Eds), *Introducing Women's Studies*. London: Macmillan.

Troyna, B., & Hatcher, R. (1992). *Racism in Children's Lives*. London: Routledge.

Walker, J. (1988). *Louts and Legends*. Sydney: Allen and Unwin.

Wallace, M. C. (1992). *West End Riots 1991: Young People's Perceptions*. Unpublished M.A. thesis, University of Northumbria.

Weiner, G. (1994). *Feminisms in Education*. Buckingham: Open University Press.

White, A. (1989). *Poles Apart?* London: Dent.

Williams, F. (1997). Feminism and Social Policy. In: V. Robinson & D. Richardson (Eds), *Introducing Women's Studies*. London: Macmillan.

Willis, P. (1977). *Learning to Labour*. Aldershot: Saxon House.

FIRST DAYS IN THE FIELD: GENDER AND SEXUALITY IN AN EVANGELICAL CHRISTIAN SCHOOL

Geoffrey Walford

INTRODUCTION

The first part of my title is rather presumptuously copied directly from Blanche Geer's classic chapter first published in 1964 – 'First days in the field'. That chapter was one of eleven included in Phillip E. Hammond's path-breaking edited collection *Sociologists at Work* which was the first to present case histories of the research process that lay behind major research projects. As Howard S. Becker is quoted as saying on the cover of the paperback edition (Hammond, 1967):

> As every researcher knows, there is more to doing research than is dreamt of in philosophies of science, and texts in methodology offer answers to only a fraction of the problems one encounters.

That edited collection was the first to present what have come to be known as 'confessional tales' (van Maanen, 1988) about the sociological research process.

Geer's (1964) contribution is about an educational research project. With such famous names as Howard S. Becker, Anselm L. Strauss and Everett Hughes, she had just finished working on the project that had led to *Boys in White* (Becker et al., 1961) and was setting out on a new, but related, project on the culture of college life that culminated in *Making the Grade* (Becker et al., 1968). In her chapter, Geer examines the fieldnotes that she took in the first

Genders and Sexualities in Educational Ethnography, Volume 3, pages 111–124.
ISBN: 0-7623-0738-2

eight days of fieldwork among new undergraduates at the University of Kansas who were attending introductory 'preview' days. Her concern is with the relationships between initial fieldwork experiences, her thinking before entering the field and her final understandings achieved at the end of the research process. She shows the way that strategies and concepts developed before entry to the field can quickly change during the initial stages of fieldwork, and how some tentative hypotheses that developed in the first days became major themes in the subsequent research. She argues that, while early fieldwork reaches few conclusions, it may nevertheless have far-reaching effects on the rest of the research.

This paper is concerned with my own 'first days in the field' on a new research project. It considers the potential impact of early fieldwork on the larger research project and also discusses the ways in which past experiences and transient personal events may influence what the observer sees and records.

ENTERING THE FIELD

The research project with which this paper is concerned is a comparative study of policy on schooling for religious minorities. The three-year project has the aim of examining policy formulation and implementation within England and The Netherlands, looking at national, local and school level policy and practice. My previous research on sponsored grant-maintained schools led me to choose evangelical Christianity and Islam as the two minority religious groups to be studied (Walford, 1997, 1998a). In order to understand the details of policy at the school level the research involves conducting 'compressed ethnographies' (Walford, 1991, p. 91) in several schools in each of the two countries. Here, over the three-year period of the research, the aim is to spend about one month, in several separate blocks, in four schools in each country.

As my recent research has been concerned with policy at the government and legislative level, it was some considerable time since I had conducted any ethnographic work. In fact, the work for *City Technology College* (Walford & Miller, 1991), where I researched the experiences of staff and students in the first of these new schools, was the last real ethnographic study. Everything since that time had been based on very short visits and interviews.

My earlier work on evangelical Christian schools (Poyntz & Walford, 1994; Walford, 1995) had centred on a group of private Christian schools which shared an ideology of Biblically-based evangelical Christianity. This work had involved a questionnaire to schools and interviews with headteachers and some others. I had not spent any time in the schools observing classroom activity.

This previous research had found that these schools were usually poorly funded, having been set up by parents or a church group to deal with a growing dissatisfaction with what was seen as the increased secularism of the great majority of schools. The schools aimed to provide a distinctive Christian approach to every part of school life and the curriculum and, in most cases, parents had an important role in the management and organization of the schools.

About 65 of these schools came together through mutual recognition into a loose grouping through the Christian Schools Trust (CST). As the number of new Christian schools increased during the 1980s, several of the heads of the schools began to meet together regularly for Christian fellowship and to discuss matters of mutual interest. More formal meetings and some conferences began to be held, and other teaching staff became involved so that, in 1988, a decision was made to establish the Christian Schools Trust "to promote and assist in the founding of further schools" (CST, 1988). The Trust provides assistance in the development of curriculum materials, helps co-ordinate the dissemination of such materials, provides some in-service training for teachers and organizes conferences. The nature of the schools involved with the Christian Schools Trust has been described in some detail elsewhere (O'Keeffe, 1992; Walford, 1994b, c; Poyntz & Walford, 1994). It is worth noting that these schools do not serve the 'traditional' private school market. Many of the schools have progressive fee structures that are linked to ability of parents to pay, and wish to be open to children from a wide social range.

This previous work meant that I had some idea of the diversity of schools connected to the Christian Schools Trust. In order to draw comparisons with Dutch Christian schools which are fully funded by the state, I wanted to undertake a compressed ethnography in two British Christian schools that I believed might well be able to able to enter the state-maintained system if they wished to do so. The schools thus had to be of a reasonable size and have buildings and facilities in which the National Curriculum could be taught with, perhaps, only minor changes. I thought that The Lord's School (a pseudonym) might be a possibility and I telephoned the headteacher to arrange an interview. He was willing to be interviewed and I spent about two hours with him – first in his office and then on a tour of the school. On the tour I was introduced to several teachers and had the chance for quick conversations with several of the children. I recognised that the school was, indeed, one that met my criteria for further study and decided to ask to be allowed to visit for a more extended time. At that point I asked for permission to stay for a week. The headteacher asked me to write explaining exactly what I wanted to do so that the staff could consider the proposal. They agreed to allow me to conduct the research in the school.

At the time of my first visit to the school there were about 140 students altogether. Classes are small, the largest having about 20 students and some below 10. Most of the teachers are qualified to teach in state schools and most have done so – often for many years – before moving to this school. However, they all receive far smaller salaries than they would in the state sector as payment of all staff is 'according to need'. This is related to the school's policy on school fees where, although there is an official set termly charge, parents are actually charged according to their ability to pay. In addition to a core of full-time staff, the school has many part-time staff, some of whom come into the school just a few times each week. Some of both groups of teachers are parents of students.

My fundamental interest in doing fieldwork in the school was simply to enable me to understand the culture of the school and the constraints under which it operated. If I were to be able to ask sensible questions about policy at the macro-level, I needed to understand what was important to the schools. However, by the time I entered the school, I realized that I should consider writing a paper about these initial days in the field. Indeed, I almost felt obliged to do so as I had recently written a chapter on publishing research where I had argued that "My writing, and to a great extent the conduct of the research itself, is always structured around particular books and journal articles that I wish to write" (Walford, 1998c). If I was going to do this period of observation, it might just as well be useful in several ways. As I was at the early stages in the process of editing this volume on *Genders and Sexualities,* it was an obvious potential focus. I felt that this particular evangelical Christian school might enable me to say something of interest about the topic. Gender issues were thus one of several 'foreshadowed problems' (Malinowski, 1922, p. 8) that I had in mind when I started these 'first days in the field'.

THE GENDER AGENDA

In selecting gender and sexuality as areas for special consideration, I already had a reasonable knowledge of the academic literature. I had taught sociology of education for many years and had an extensive booklist on the subject of gender. Although my own writing on gender is limited, I had just completed co-editing a volume on *Children Learning in Context* (Walford & Massey, 1998) that contained several ethnographic studies examining aspects of gender in primary and other schools. I had a less extensive background knowledge of the academic material on sexualities, but had read a reasonable amount. I knew what sort of activities would be worth trying to observe and comment on.

The first days in the field are often seen as the most challenging and emotionally awkward. Meeting any new group of people in an environment which they already inhabit can be uncomfortable and embarrassing, but it can be particularly so where those being met are to be research 'subjects' and do not fully understand the nature of ethnographic research. Yet the first days of research are also often seen as particularly exciting, for so much of what is experienced and observed is new to the researcher. Indeed, the researcher is often overwhelmed by the amount of new information that it is necessary to take in. Bogdan & Taylor (1975) advise limiting observation periods at first to an hour or less, and spending a great deal of time writing fieldnotes. They remind us that "Observations are only useful to the extent that they can be remembered and accurately recorded" (p. 42). In contrast, Geer (1964) recorded all her notes at the end of each full day. I followed the latter's example and typed up notes on my laptop at the end of each day.

In taking the decision to write notes at the end of each day, I was trying to balance the ideal with what I felt was practically possible. Whilst it might have been highly desirable to have made notes during the actual lessons, I felt that, at least in the initial stages, this might be too disconcerting for the teachers involved. On this first visit I wanted to try to encourage these teachers to trust me and to feel reasonably at ease with me in their classrooms. I wanted to show them that I was 'no danger' to them. The problem with this decision, of course, is that these summative notes might be coloured by 'generalized' impressions of the day. Important material might be forgotten or misremembered, and particular incidents might be influenced by the overall impressions of the day. While I tried to avoid these problems, and I did make some short notes at odd times during the day in an attempt to remember important details, I cannot claim that I completely overcame the difficulty.

Given that I had gender and sexuality as a focus for my observations, it was inevitable that many of the notes I made would relate to that issue. However, I was immediately struck by just how much there was to record. From the way that some of the classes lined up to enter or exit rooms, to the boys' football in the playground which occupied nearly all of the main space, gender differentiation and stereotypical gender reinforcement were very evident. Girls were clearly dealt with in a more gentle way than boys, and boys were expected to be tougher and 'louder' than girls. But I also soon recognized that the ways in which the school dealt with gender and sexuality issues were not straightforward. While the reinforcement of some aspects of heterosexual gender roles was seen as central to the values of the school, other aspects of 'traditional' gender roles were challenged.

It is important to recognize that, while the fundamental reasons for the establishment of Christian schools relate directly to the perceived secularism of state-maintained schools (Walford, 1994b), issues of gender and sexuality are far from incidental to their existence. The secularism with which parents are concerned has many facets, but the liberal view of sexuality that is believed to be propagated in many state-maintained schools is a key issue. For these Christian parents, sexuality is a gift from God that should not be abused. It should be reserved for heterosexual couples within marriage. Early 'experimentation' with sex before marriage is seen as sinful – as is any homosexual activity.

These strongly held beliefs are felt by the teachers in this school to be compromised by the liberal form of sex education given in many state-maintained schools and with what is perceived as the acceptance of the validity of diverse 'lifestyles'. This concern about homosexuality was voiced explicitly on several occasions during my first days. For example, during break-time one teacher was talking about why the school did not want state support. He said that they did not wish to be 'politically correct' in their teaching of sexuality. It was clear from further discussion that he meant that they would not teach homosexuality as a 'valid lifestyle'. He showed confusion about the effects of Section 28 of the Local Government Act of 1988 believing that it actually encouraged the promotion of homosexuality as an equally valid lifestyle (when it actually bans the 'promotion' of homosexuality by local authorities) and that it applied to state-maintained schools (when this has been the responsibility of individual schools since 1986) (see Epstein & Johnson, 1997, p. 14). So I explained to him that, although many people believed this, it was not the case and that sex education was under the control of the Governing Body of each school. It is only within science that the 'mechanics' of some forms of sexual reproduction have to be taught at a set age in accordance with the National Curriculum. Nonetheless, he persisted that to accept state funding might compromise the school in the future – further legal requirements might be introduced that forced them to go against their principles.

With this strongly anti-homosexuality belief being held by (I assume) all the staff, I was surprised how often issues related to homosexuality were evident during these first days of observation. For example, in geography a film about the effects of tourism in Sierra Leone was shown. Men were shown dressed as women seductively dancing a 'Hey Big Spender' routine for tourists. The teacher commented how sad this was, but did not discuss explicitly any potential homosexual or prostitutional meanings. Not so for the RE teacher who unhesitatingly categorized members of 'the gay community' as being amongst 'prostitutes and other sinners'.

Another teacher expresses this concern with sex education in terms of it being one of the areas of difference between the school and the state system. She said that the school wanted the freedom to do sex education when a particular group was ready, not when it came on the syllabus. She believed that some groups of children were ready for sex education early, while it would be better to delay it for other groups. Also crucial was that they wished to teach sex education only within a Christian context which emphasized that any sexual activity outside marriage was sinful.

I found it interesting that these conversations about sexuality were often accompanied by ones on corporal punishment. Both of the teachers discussed above also mentioned corporal punishment during the same discussion. I am not implying here that the teachers saw any direct link between sexuality and corporal punishment, but simply that these two areas were seen as areas where the school's policy was opposed to that of state-maintained schools and also opposed to the prevailing 'politically correct' view. Corporal punishment was only very rarely given at the school, but all parents were required to sign a form giving permission for corporal punishment to be used should the school believe it necessary. It was related directly to the belief that Christians should "train up a child the way he should go" and that "parents should discipline their children" (see Walford, 1995, ch. 3). This was seen as part of the 'Christian ethos' that they would have to give up if the school became funded by the state. In fact, there was similar confusion exhibited about this issue as about the issue of sex education. The 1998 Education Standards and Framework Act had already made corporal punishment illegal in all schools (including private schools) before the fieldwork period. One of these teachers did not know how this had 'turned out' and talked about a parent from another school going to the European Court to get the right for schools to use corporal punishment.

Although I knew that 'appropriate' gender and sexual behaviour was important for these schools, I was still surprised by the number and nature of the events that I recorded that attempted to help shape the gender roles of students. Perhaps one of the most obvious aspects (but one that it is easy to overlook) is the different uniforms that boys and girls wear. The school has a uniform code which is strongly enforced. Girls wear dark pleated skirts, black tights, very 'sensible' black shoes and a dark blue pullover. The whole effect is exceedingly drab. They also wear school ties on white shirts. Boys have black shoes, dark trousers, dark blue pullover, white shirt and a school tie. The essential difference is that the girls are never allowed to wear trousers except during games periods. It was very hot on one of the first days I visited which led to an incident which gave an insight into the perceived importance of sex-appropriate dress. One boy took off his pullover and tied the two arms round his waist and let the body of the pullover hang behind him. A teacher told him off for doing so, saying that he should not be 'wearing half a skirt'.

What was particularly noticeable was that the type of crude gender differentiation that was found by such researchers as Delamont (1990), Evans (1988), Measor & Sikes (1992), Stanworth (1983) or Wolpe (1988) was still echoed in this school. But here, it was not seen as problematic nor still occurring because the staff were not up-to-date, but was seen as a sensible and desirable way of structuring and maintaining gender identities. Thus I observe registers being called with the names of boys and girls separately listed (either by Christian names or by numbers with girls first, followed by boys). Gender was also reinforced by girls and boys being lined up separately when moving from room to room as a group. None of this was done rigidly, and not all teachers did it, but it was sufficiently present for it to be accepted as common practice. I was also told that in a previous year, to deal with an unusually large year group, the school had decided to have a separate boys' class and a girls' class. One teacher explained, "We wanted the boys to gel as a group and the girls to gel as a group, and that has happened."

In the classes I observed, all the children appeared to have been able to choose their own seating positions. In most cases this meant that boys tended to sit with boys and girls with girls, but there were many overlaps. Children were sometimes moved by the teacher to another seat. Where this was done, there was a tendency for it to be a move away from a single gender group to a dual gender group.

Stereotypical gender behaviour was left unchallenged in many of the classes I observed. For example, in a computer lesson some of the girls deferred to the 'superior knowledge' of the boys about computing, saying that "the boys are much better" at computing. In English a free choice was given for a small research project. It was not difficult to guess who chose cars, war or monsters, and who chose hairdressing or the family. Of course, there were also many topics that might equally well have been chosen by boys or girls, and some of the girls in the computer class actually had far greater knowledge and skill than some of the boys, but it was noteworthy that it appeared to be broadly accepted by boys, girls and the teacher that some topics were more appropriate for particular genders.

However, in contrast, some aspects of gender stereotyping were also challenged by the school or, at least, were more ambiguous in their effect. Games, for example, are taught in mixed gender groups. This means that students play in mixed gender groups and all are expected to take an equal part.

In practice, however, I heard complaints from the boys when one side had 'more boys' than the other, and the teacher took some care to ensure that the groups were roughly equally balanced in terms of boys and girls. This female teacher was also accepting of the idea that one boy in particular had found it difficult having a woman teaching games. It was certainly true that he had loudly and rather rudely complained about her judgements.

One important example where gender stereotypes were challenged related to the way that the children were required to look after their own classrooms. An unusual feature of the school was that the children did the everyday cleaning and clearing-up at the end of the school day. This was partly done to save money as the school was poor, but also partly done to ensure that the children developed a considerate attitude towards others and shared in the manual work required to keep the school clean. But, rather than have the boys do the 'heavy lifting' and the girls do the 'light dusting', both boys and girls were required to engage in a full range of activities. And none seemed to object.

This attitude of consideration for others was evident in both boys and girls. They were not, of course, without fault, but there appeared to be a high level of support for each other and acceptance of each other's differences. This was not just a matter of the lack of any observable bullying or teasing of each other, but that they showed respect for different understandings and different levels of ability. Several of the children had moved into the school following bullying at a state-maintained school. Here they were able to find a safe environment where they could be different. It was possible for a boy to be gentle, or to not like sport. In a school where going to church is accepted, where children sometimes pray for each other in class, where both male and female teachers show a real personal care for children in their class, where miracles are believed to occur, not liking sport is a trivial failing for a boy.

GENDER, SEXUALITY AND STAFF

The importance of gender and sexuality extended to staff as well as students. Only Christian staff are appointed to the school, and all have to agree with the Christian ethos of the school. In practice, this presents little difficulty as there is no set salary structure and staff have to be prepared to receive a salary 'according to need'. It is thus highly unlikely that any teacher or other staff member would wish to work at the school unless they fully supported the evangelical Christian orientation of the school. This means that staff are likely to agree with a particular view of the importance of sexuality and gender within their world-view.

This is exemplified through the general view abut the need to have both male and female staff. While there are more female than male teachers, the school has a significant number of both, for the school believes that it is important that boys do not only have female teachers. I was told that for one recent vacancy the staff had specifically prayed for a male teacher. I was told that, very soon, without any advertising or that person having any knowledge of any vacancy, a male teacher applied to the school to see if they needed help. This man was appointed.

The importance of being able to select Christian staff within the same theological tradition was a further reason why the school did not wish to seek state funding. As homosexuality was viewed as sinful, homosexual activity was regarded as incompatible with being a teacher. One teacher told me that the school would, "of course, never appoint a homosexual teacher." Indeed, he suggested that the fact that there were homosexual teachers in state-maintained schools was one of the reasons some parents used this school.

One further feature relating to gender and sexuality that I observed was that, while the staff generally dressed rather informally, there was also a clear 'dress code'. One part of this is rigidly enforced: that women should not wear trousers. They are allowed to do so when doing games or PE, but this is very much seen as an exception. What is interesting here is that, when I asked about this, some of the female teachers made it clear that they would much prefer to be able to wear trousers. They considered them more practical, but they were forced to wear skirts and dresses against their will. This particular restriction on dress is also found in some state-maintained and in other private schools, but here the pressure is difficult to resist because it links to particular views of femininity that are found within the evangelical Christian church.

I also recorded a high degree of camaraderie between the male teachers. For example, in one semi-public discussion in the staff room it became evident that the men regularly go for a meal together once or twice a term. This was "not sexist," one of the men quipped, it is "just restricted to those who have beards." If there were any "women who had beards," then they could come too.

A further feature about staffing is that the headteacher was male. This is not a chance feature of the particular school, but links to a finding of some of my previous research that several of these evangelical Christian schools express the belief that women should not take up 'leadership positions' in the church or in activities related to the church (Walford, 1995). Basing their view on the Genesis instruction that man should have dominion over women, they believe that it would be wrong to have male teachers under the authority of a female headteacher.

MULTIPLE AGENDAS?

There are two aspects of this account of 'first days in the field' that I feel are worth reflecting on further. The first is methodological and the second substantive.

First, my notes about the gender and sexuality became less frequent as the observation period progressed. I made very many negative notes about gender stereotyping and gender role socialization in the first couple of days, but far fewer at the end of the week. My notes on the early days were largely negative in tone but, in contrast, my notes towards the end of the week contained far more positive descriptions. I still had many negative comments, but I also had far more accounts of what I interpreted as positive activities and events. For example, one note read:

> Perhaps the most important thing about the day is to record the feeling of the place. This felt like a place with 'real' Christians. They were very friendly towards each other. There was a clear belief system. Teachers could walk into each other's rooms, and it felt OK. They seemed to have a common aim. Academic standards seem to be good, but it is not taken too seriously as a fundamental aim. The science teacher said to me that he thought that much of the science GCSE was a waste of time. They had to teach it because of the exams, but that he wanted to go beyond that.

My concern is that I am unsure about the extent to which these differences are due to the teachers and how they conducted themselves on the particular days I observed them, or to changes in the way I observed and interpreted what it is to be a child and a teacher in that school. To what extent were the differences due to my changing perception and interpretation of events?

Now it could be that, after the first couple of days, I had simply exhausted the comments that could be made about the aspects concerned with gender and sexuality. But, as an ethnographer, I should still be recording incidents that illustrate such activities. I should not stop recording simply because I have already recorded a similar incident. However, if the culture of the school and the everyday behaviour of the teachers and children are relatively fixed, it is ridiculous to expect an ethnographer to record daily that, for example, the register was called with girls first, then boys. At the very least, this highlights that the number of recorded incidents or cases gives no indication of the significance of a particular activity.

But it may be that my change in what I recorded was not due to a conscious decision not to repeat recording the same event. It may be that I was simply getting used to the school's culture and no longer seeing it as 'strange'. In her guide to research, *Fieldwork in Educational Settings*, Delamont (1992, p. 99) gives an example where she was shocked by the way in which children with special disabilities were treated by teachers, and the way in which those children accepted the behaviour and designations of teachers. She argues that experienced teachers of children with learning difficulties would probably have found such treatment 'normal' and so, presumably, would the children. It may only be in the early days of fieldwork that a researcher is able to problematize the 'normal' and begin to ask questions about the ways in which the participants in any culture structure their behaviour, beliefs and meanings. The pressures on researchers to be accepted by those in a new culture is such that this period of 'culture shock' may be very brief. In Delamont's (1992, p. 97) terms, it is easy to sink into the 'warm sand'. There is a need to maximize the research insights that can be gained while the researcher is still 'alien'.

It may be that ethnography within schools is even more difficult than most educational ethnographers admit. While it is possible for the ethnographer of the New York Stock Exchange (Abolafia, 1998) or of sexual behaviour in public toilets (Humphries, 1970) to enter that world with very little previous knowledge, the vast majority of ethnographers of schooling have experienced at least twelve years within schools as students and many such ethnographers have experienced many more as teachers. This very long period of socialization into the culture of schools cannot easily be negated or the understandings of what is 'normal' in such cultures problematized. My feeling now is that at least some of the differences between the fieldnotes that I made on the first days and those at the end of the week were due to my rapid re-acclimatization into a culture that I had lived through in my formative years. If the Jesuits are correct, the first few years of life are crucial. Perhaps I slipped back into this culture very quickly and I began to see the school's processes of gender socialization of children as 'normal'.

Second, there were several cases where it seemed that the activities of the school were constrained by the most conservative elements associated with the school. One of the most obvious examples was the fact that female staff were not allowed to wear trousers, as mentioned above. It was evident that some of the women could see no religious or other reason for this rule, but that they were unable to make changes because this was one of the aspects that marked off the school as 'different' by some of the parents. In a similar way, some teachers believed that there should be the possibility of having a female headteacher. But, as some of the parents and teachers believed that this would be unscriptural, those who were more liberal in their interpretation were prepared to accept the more conservative viewpoint.

As with other fee-paying schools, Christian schools live in a real market. For many of these parents, taking their children out of the state sector of schooling is a major and unprecedented decision. Most of these parents are not traditional users of the private sector, but are looking for a particular form of schooling that they feel is appropriate for their children. What I began to see in these first few days of fieldwork is that the more conservative parents probably have a greater influence on the running of these schools than do the more liberal parents. If some parents object to female teachers wearing trousers, then it is probably easier for the school to acquiesce to their demands than to make a point of not doing so. However, the teachers' acquiescence to the views of some were not simply a response to market pressures, but were linked to a belief amongst some of the teachers that one should not challenge aspects of someone else's faith. Indeed, this might be seen as part of a Christian ethos where one is compliant to the wishes of others. Likewise, if, on principle, some parents do not have television or video at home, the school is likely to be very sparing in using them in teaching. The influence of the more conservative parents and teachers may be greater where the school is in a real market which involves the payment of fees than in a quasi-market that does not.

Here is an important new idea that may develop into a theme in the larger comparative project between England and the Netherlands. As the Dutch Christian and Muslim schools are fully funded by the state, they present a contrast to those in England that have to rely on fees. One might expect these Dutch schools to feel less compliant towards a minority of parents who have extreme views. But an investigation of the extent to which that is true goes far beyond what can be gained during these 'first days in the field'.

ACKNOWLEDGEMENTS

I am most grateful to the head and teachers of the school for allowing me to observe. The research reported in this paper was made possible by a grant from the Spencer Foundation. The data presented, the statements made and the views expressed are solely the responsibility of the author.

REFERENCES

Abolafia, M. Y. (1998). Markets as cultures: an Ethnographic approach. In: M. Callon (Ed.), *The Law of the Markets*. Oxford: Blackwell.

Becker, H. S., Geer, B., Strauss, A. L., & Hughes, E. C. (1961). *Boys in White: Student Culture in Medical School*. Chicago: University of Chicago Press.

Becker, H. S., Geer, B., & Hughes, E. C. (1968). *Making the Grade: The Academic Side of College Life*. New York: John Wiley.

Bogdan, R., & Taylor, S. J. (1975). *Introduction to Qualitative Research Methods*. New York: John Wiley.

Christian Schools Trust. (1988). *Information Sheet.*

Delamont, S. (1990). *Sex Roles and the School* (2nd ed.). London: Routledge.

Delamont, S. (1992). *Fieldwork in Educational Settings.* London and Washington, D.C.: Falmer.

Evans, T. (1988). *A Gender Agenda.* Wellington and London: Allen and Unwin.

Epstein, D., & Johnson, R. (Eds) (1997). *Schooling Sexualities.* Buckingham: Open University Press.

Geer, B. (1964). First days in the field. In: P. E. Hammond (Ed.), *Sociologists at Work.* New York: Basic Books.

Gribble, D. (1988). *Real Education. Varieties of Freedom.* Bristol: Libertarian Education.

Hammond, P. E. (Ed.) (1964). *Sociologists at Work.* New York: Basic Books.

Hammond, P. E. (Ed.) (1967). *Sociologists at Work.* New York: Doubleday Anchor.

Humphries, L. (1970). *Tearoom Trade.* London: Duckworth.

Malinowski, B. (1922). *Argonauts of the Western Pacific.* London: Routledge and Kegan Paul.

Measor, L., & Sikes, P. (1992). *Gender and Schools.* London: Cassell.

O'Keeffe, B. (1992). A Look at the Christian Schools Movement. In: B. Watson (Ed.), *Priorities in Religious Education.* London: Falmer.

Poyntz, C., & Walford, G. (1994). The New Christian Schools: A Survey. *Educational Studies, 20*(1), 127–143.

Stanworth, M. (1983). *Gender and Schooling.* London: Hutchinson.

van Maanen, J. (1988). *Tales of the Field.* Chicago: University of Chicago Press.

Walford, G. (1991). Researching the City Technology College, Kingshurst. In: G. Walford (Ed.), *Doing Educational Research.* London: Routledge.

Walford, G. (1994a). *Choice and Equity in Education.* London: Cassell.

Walford, G. (1994b). Weak Choice, Strong Choice and the New Christian Schools. In: J. M. Halstead (Ed.), *Parental Choice and Education.* London: Kogan Page.

Walford, G. (1994c). The New Religious Grant-maintained Schools. *Educational Management and Administration, 22*(2), 123–130.

Walford, G. (1995). *Educational Politics: Pressure Groups and Faith-based Schools.* Aldershot: Avebury.

Walford, G. (1997). Sponsored Grant-maintained Schools: Extending the Franchise? *Oxford Review of Education, 23*(1), 31–44.

Walford, G. (1998a). Reading and Writing the Small Print: The Fate of Sponsored Grant-maintained Schools. *Educational Studies, 24*(2), 241–257.

Walford, G. (Ed.) (1998b). *Doing Research about Education.* London and Washington D.C.: Falmer.

Walford, G. (1998c). Compulsive Writing Behaviour: Getting it Published. In: G. Walford (Ed.), *Doing Research about Education.* London and Washington D.C.: Falmer.

Walford, G., & Massey, A. (Eds) (1998). Children Learning in Context. *Studies in Educational Ethnography, 1.* Stamford, CT and London: JAI Press.

Walford, G., & Miller, H. (1991). *City Technology College.* Buckingham: Open University Press.

Wolpe, A. M. (1988). *Within School Walls.* London: Routledge.

FLIRTING FROM THE THRESHOLD: ESCAPING THE GENDERED DIVISION OF LABOUR THROUGH SEXUAL AMBIGUITY – A CASE STUDY OF LESBIAN OFFICERS IN THE SEA CADET CORPS

Jayne Raisborough

INTRODUCTION

It'd be easier sometimes. Its like a drill squad, They march as one, one solid block. A whole. But I'm out, make myself out. I've got to because I don't like what's happening. I'm like a spanner in the works. *Fay*

It's like they can't place me anywhere. Maybe I'm gay, maybe I'm just a feminist, maybe just intelligent. I don't fit and they just can't handle it. *Michelle*

These quotations illustrate the feelings of difference, or lack of 'fit' experienced by (closeted) lesbian officers in a British uniformed youth organization, the Sea Cadet Corps (SCC). Michelle feels that she cannot be 'placed' by others in the SCC and Fay feels like a 'spanner in the works'. This chapter explores how lesbian officers maintain an ambiguity that disguises their lesbianism but which challenges normative notions of heterosexual femininity which underpin a gendered division of labour in the SCC. Lesbian officers can maintain their ambiguity because of the contradictory discourses currently

Genders and Sexualities in Educational Ethnography, Volume 3, pages 125–139.
Copyright © 2000 by Elsevier Science Inc.
All rights of reproduction in any form reserved.
ISBN: 0-7623-0738-2

circulating the SCC around the issue of homosexuality. Discourses of political correctness that compete with those of institutionalized homophobia, allow lesbian officers to move onto the *threshold* of the closet. From the threshold lesbian officers perform an ambiguity which problematizes their positioning in a gendered division of labour. The threshold is also an empowering place to disarticulate homophobic and sexist voices.

THE SEA CADET CORPS

The SCC is one of three uniformed youth organizations in the British Isles, which earn financial assistance from the Ministry Of Defence (MOD). The main objective of uniformed youth organizations is to promote responsible citizenship in young people through military-style training and discipline while offering them an insight into a possible military career. The SCC is the smallest of the MOD-assisted youth organizations with 400 units in the British Isles. Each unit has a complement of volunteer adult staff (aged 18–55) and cadets (aged 10–18). The SCC has a strong relationship with its parent organization, the Royal Navy (RN). All members wear RN uniforms, train on RN ships and bases and follow a training curriculum which is loosely based upon the RN's. Both staff and cadets progress through a career structure of RN promotions as they earn SCC qualifications in a range of skills: from engineering, cooks, stewards, meteorology, radio communications and stores management, to shooting and water sports.

My study of the SCC is ethnographic, involving participant observation, interviews, snatched conversations and 'hanging around'. I have been in the field for over four years. At first my field had a physical location as I focused upon one particular SCC unit called TS Response. However the field quickly expanded to incorporate residential training courses on Naval Bases, adventure camps, and other units. This expansion was mainly due to the efforts and intervention of the women of TS Response and to the friendships I had made during my own membership of the SCC. The women of TS Response introduced me to women of other units, and a long standing friend (now a senior officer of the SCC) used her position to enable my access to further units and courses.

The data for this chapter are only partially drawn from my observations over different sites. My main data come from in-depth conversations I had with three women who during the course of this fieldwork became close friends. The three women, unknown to each other, all identified as lesbian and defined their sexuality as closeted – that is to say that none of the women had revealed their sexuality to anyone in the SCC (although one of the women was 'out' in her workplace). I only had the opportunity to observe one of these women as she carried out her duties or relaxed in the SCC environment. The majority of what is reported here comes from the women's own narratives and an analysis that brewed during our conversations.

My wider research explores the experiences of women in the SCC, examining the discursive and material restrictions upon their participation. I argue that women are expected and encouraged into labours and activities which accord with dominant notions of femininity. There is, then, as in other work and leisure organizations, a gendered division of labour in the SCC. My research analyses how this division of labour discursively reproduces and invokes dominant notions of femininity articulated through institutionalized heterosexuality. I am interested in how these processes operate to render women's participation in the SCC both limited and conditional.

Lesbian officers constitute an illuminative case study as they form the small number of powerful women in the hierarchy of the SCC. In short the lesbians of this study are women who have 'broken out' from the normal restrictions placed upon women and are able to position themselves outside the gendered division of labour. This is not to suggest that the only powerful female leaders in the SCC are lesbians. Of the five women interviewed who held positions of authority, three revealed their lesbianism during the course of the research. Of the three women who make up this case study,[1] two initially stated that their sexuality had no impact upon their 'job' at the SCC. Yet as we explored this closely, we found that lesbians, although not 'out' because of fear of homophobic retaliation, were not completely closeted and used their position on the threshold of the closet for political effect.

LESBIANS IN ORGANIZATIONS

Research on identity management records how lesbians undertake self-surveillance and self-censorship to control or avoid the possible revelation of their sexuality (Clarke, 1996; Levine & Leonard, 1984; Sparkes, 1994). In order to negotiate hegemonic heterosexuality in organizations, lesbians are argued to perform a pseudo-heterosexual identity which passes the fierce, subtle forces of homophobia. As part of this identity lesbians employ a number of performative strategies, from the creation of necessary fictions pertaining to male partners, to avoiding personal disclosure, to performing overt heterosexuality by flirting with men. These 'props' signify their sexuality as normative and conceal their lesbianism. While constant performances of concealment are exhausting and ridden with angst, some lesbians find them amusing, empowering and even titillating. As one of Gill Clarke's lesbian PE teachers states, "it's quite exciting in a way getting away with it" (Clarke, 1996, p. 205).

In contrast to the lesbians who conceal their sexuality through consistent performances which allow them to pass, lesbian officers in the SCC take great delight in performing an ambiguity. Fay says that:

> No [I'm] not out, never out but I'm not exactly in either. I'm teasing them – really daring them.

As a senior officer in a the SCC, Fay is neither 'out' nor 'in'. She creates a performance not of pseudo-heterosexuality (passing as straight) nor of lesbianism, but one which is ambiguous and one which reveals the artifice of this ambiguity: she is 'teasing'. Despite experiencing the SCC as homophobic, Fay and the other lesbian officers, Michelle and Debbie, are able to perform a *sexual ambiguity*. The performance of ambiguity involves inconsistent performances of sexual and gender identities.

> Yeh, I hop from one foot to another, I can do the girly bit, I do the straight stuff, then the gay stuff. It all gets jumbled up and you can see them thinking 'where is she coming from?' *Debbie*

> On the whole it puts them off guard, I can go along with so much, it lulls them into a false sense of security, and then I'll just chuck something in, like 'partner' or mention *Ellen* or when I said, 'Oh yeah, I've seen *In and Out*' and they're thinking, "Why'd you go and see a film about a gay teacher?" *Michelle*

While sending what Debbie calls 'gay signals', lesbian performances of ambiguity also allow officers the safety of the closet. While littering their conversations with references to 'partners', gay stars (e.g. Ellen) and gay bars, they openly flirted with men, talked about their hopes for motherhood[2] and joined in both 'men's' and 'girls' talk when sexually objectifying members of the opposite sex. Debbie illustrates how she associates herself with a homosexual lifestyle, but does not publicly identify with it.

> It's trendy to go to Gay bars. So they say, "You went there?" disgusted. And [I'll] say, "What the hell do you think happened, what you think it's like? They're not perverts you know." *Debbie*

Why and how performances of ambiguity are possible within the SCC organization is explored here. As organizational discourses are important in the shaping of normative engendered sexual identities, the next section examines the contradictory discourses which constitute the SCC's approach to homosexuality. I will argue that the existence of pseudo-tolerance towards homosexuality enables lesbian officers to position themselves upon the threshold of the closet. Here they still retain some of the safety of the closet, but have a better vantage place to challenge homophobia.

THE SEA CADETS: HOMOPHOBIC POLITICAL CORRECTNESS AND 'PSEUDO' TOLERANCE

Despite the strong relationship between the SCC and the RN, the SCC falls under the remit of civilian and not military law (see Hall, 1995). Therefore the SCC has no official policy that excludes lesbians and gays. Indeed, when potential SCC officers are trained, they are told that they cannot dismiss or exclude a member of staff, or a cadet on the grounds of sexuality. However, this message often follows a problem solving exercise, where officer recruits are asked to respond to imaginary scenarios involving two adult male staff having sex, or an adult man kissing a male cadet.[3] These scenarios reflect that the SCC's organizational view of homosexuality is primarily sexual, male defined, secretive, with additional associations with child abuse.[4]

The SCC has a contradictory stance towards homosexuality, it is torn between the RN's stance of homosexuality (i.e. exclusion), and the need for the corps to become more inclusive and representative of its potential recruitment pool.[5] Some management personnel regard this as the latest wave of political correctness to hit the SCC.[6] Many dismiss it, other senior managers are more circumspect when airing 'un-PC' comments, often retreating or explaining their position when challenged. At best the management of the SCC adopts a pseudo-tolerance. This is a liberal stance which accepts the existence of homosexuality but keeps it firmly in the private domain. Homosexuality is a private (presumably, sexual) act best left to the discretion of individuals in the privacy of their homes. As Smith (1995, cited in Clarke, 1996, p. 194) remarks, pseudo-tolerance is reserved for the 'good' homosexual, one who maintains the division between their private homosexuality and their public performances of heterosexuality. The 'good' homosexual is therefore invisible.

Debbie argues that the 'good' homosexual is not just invisible in the SCC, but rendered non-existent. She claims that pseudo-tolerance works on the basis that all personnel are heterosexual and that the issue of homosexuality will not arise:

> It's like they're sending these signals that it's OK to be gay in the cadets, but they speak as if that's never going to happen, gays aren't in the cadets. That they'd be able to spot them anyway. God knows what would happen if we all stood up and said "we're gay." It's the same, it's like a bunch of white people saying we're not racist, but how do they know if there's no Blacks in the cadets? Slapping each other on the back saying what nice guys we are, not prejudiced in our little straight world.

Pseudo-tolerance then, not only renders lesbian and gay personnel invisible, but assumes that this invisibility has an equivalence with non-existence. As Debbie indicates, lesbian officers are aware of the limitations of pseudo-tolerance.

Each officer knew the SCC's pseudo-tolerance could never extend to the visible or known: that pseudo-tolerance of the invisible could not transform into acceptance of the visible.

Managerial discourses then falter along the traditional naval stance, political correctness and pseudo-tolerance. Amongst this discursive mélange are the (confused) homophobic views of some senior managers:

> One senior officer said that the world would be a better place if everyone was gay. What the hell does that mean? This, like, follows, the old "But I wouldn't want to be in a shower with one." *Fay*

If the 'official' line is somewhat clouded, the general cultural attitude is more explicit. Gay and lesbian personnel perceive and experience the SCC as homophobic, and sexist. Indeed as an observer I have often heard such terms as 'fag', 'dyke', 'pouf', 'shirtlifter' and 'lessie'[7] used to police behaviour of all members of the SCC. It is the interesting discursive mix of a confused tolerance with an attempt towards political correctness that provides lesbian officers with the space to challenge and resist organizational homophobia. They are enabled by the unique mix of these discourses to take their position on the threshold of the closet.

> [I] could be right closeted, but all this PC stuff sorts of makes me braver. I sort of want to dare them to not be PC. *Debbie*

> It's being discussed, OK in hushed tones, but it's being discussed. That makes it easier for me to chip in. All the gay issues aren't coming from me. I'm not bringing them up, I'm adding to it. *Fay*

Although it is still not 'safe' to be out, lesbian officers take advantage of the pseudo-tolerance to step onto the threshold. This action can be motivated by anger:

> [I want] to show them it angers me. The lip service. It's all the same ideas underneath but they're repackaged. Repackaged into this. *Fay*

The closet threshold is used for political effect by the three lesbian officers. It is from this position (a position supported and maintained by their performances of ambiguity) that they can highlight the hypocrisy of the SCC's pseudo-tolerance.

> PC is a load of crap really, especially in the mouths of these senior male officers. They say one thing and do something else. But then it's also something for me to use. I just have this look, or ask a simple question, "Is that right? So how does that work?" I asked some guy the other day why he felt gays shouldn't be in. He just stared. Thinking I'm one, but some might think it, but there's also some that'll swear I'm not. *Michelle*

Performances of sexual ambiguity lend weight to the lesbian officer's critiques of pseudo-tolerance. Their inconsistency allows them to raise issues which ordinarily may lead others to identity them as homosexual. Debbie argues that once she would never discuss what she describes as 'gay issues', for fear that her interest may reveal her lesbianism. In an environment of pseudo-tolerance raising 'gay issues' doesn't carry the same revelatory effect. As she has created an ambiguity around her sexuality, but has not completely identified as 'other', it is harder for other SCC members to dismiss her comments.

However, as well as challenging pseudo-tolerance, they also use the discourses of political correctness to serve their own aims. Fay uses 'political correctness' to police the homophobic language and actions of her staff:

> I let them know that it's not acceptable. They tell a joke and I don't laugh, it makes them uncomfortable. By the time we've spent any time together they've learnt not to use that language in front of the boss. My boss comes in – he's sexist and homophobic, a real bigot – but that's not the impression he gives. He is politically correct, it's OK as long as it's not in my back yard. He doesn't think there's any queers in his back yard.

> Kids shout 'fag' to each other, maybe they don't know what it means but I bollock them all the time. I say, "Why'd you call him a fag? It's not the right thing to say, it's offensive." For me it's swearing. *Debbie*

> Some kids did the 'lessie' cough when I walked past, and [I] turned round and said, "Are you casting aspersions on my sexual orientation?" I said it joking. The girls hanging around loved it, and now they say it if anyone calls them a lessie. It's like they see it as the ultimate put down. *Michelle*

The position on the threshold is supported by their performances of sexual ambiguity. Their inconsistent performances refuse their positioning as 'in' or 'out' of the closet. Michelle described her threshold positioning as one of 'flirting' She flirts with the performances which define normative (hetero)sexuality and undermines them with what she described as 'gay' performances.

> I once hugged this woman who'd done some work for me. I hugged her and practically screamed, "I could marry you and have your children."

As she flirts with different performances, she effectively flirts with those around her who struggle to position her within normative frames of reference.

> None of them would be surprised if I was gay, and no one would if I was straight. It's like things cancel other things out, like she's never got a fella, but she's one of these young women with a career, why'd she's need a man?

The threshold is valued by the three women. Each felt that they did more 'good' from this position than if they were fully out. They could never make their challenges without the safety of the closet. As their performances of ambiguity trouble the revelation of their identities as lesbians, they are allowed a stronger voice.

> Even if I got over all the nasty remarks and that, never mind what the [cadet's] parents would say, but every time I'd open my mouth, it would be that I was only saying that, doing that, because, because she's a lesbian. *Michelle*

The closet threshold is experienced as an exciting, yet dangerous place. The lesbian officers were fearful that they would not get the 'balance' right, and 'out' themselves. To be confirmed as 'lesbian' would disarticulate their anti-homophobic voices, and reduce their participation to a supposed homosexual agenda

It would be the end, I'd have to leave. *Fay*

However, it is too simplistic to assume that the threshold of the closet only enables lesbian officers to challenge homophobia. The threshold also enables lesbian officers to refuse positioning as gendered 'others' in the organization. That is to say, that performances of ambiguity from the threshold resist the positioning of lesbian officers by normative discourses of femininity, which, I argue underpins an engendered sexual division of labour in the SCC.

THE GENDERED ENCLAVE

As with other uniformed organizations, men's and women's sexuality is questioned if their gender performance does not match the socially constructed and normative ideal. Men who do not adequately perform overt heterosexuality are soon considered effeminate, while confident, assertive women are labelled lesbians (Hall, 1995; Leinen, 1993). While these terms generate differing policing practices upon individuals[8] they reflect and reproduce homophobic and normative heterosexual relations. Normative heterosexual relations involves examining dominant notions of masculinity and femininity within the SCC. It is my contention that lesbian officers perform their ambiguity not only to expose the limitations of pseudo-tolerance, but also to escape positioning as the gendered and sexual 'other' in the organization of the SCC.

Despite the policies and rhetoric of equal opportunities which couple with a generally meritocratic organization of leadership, there is a gendered and sexual division of labour within the SCC. Men are expected and encouraged into traditionally male areas of engineering and weaponry. They are encouraged to become leaders in their field, but are not so constrained by their specialized knowledge that other promotional opportunities are closed to them. SCC women predominate traditionally 'female' areas of specialization such as cooking, stores management and administration: work which is traditionally expected in the domestic, private sphere. Furthermore, SCC women rarely enter into senior management/ leadership positions, especially those which lie outside 'female' areas of speciality (cooks, etc.). So, while men can move up to become Commanding Officers of units or move into other management posts, it is still rare for women to command outside their specialized field.

My explanation for this gendered sexual division of labour in the SCC is based upon a conceptualization of the private sphere (domestic) entering the public (the SCC) as an *enclave*. The enclave is a discursively produced and policed border around women's participation in the public sphere. The enclave is maintained by bureaucratic discourses, which work to position people within the organization's hierarchy, provide them with an organization identity and police their performance (Ferguson, 1984; Holmer-Nadesan, 1996) The key here is the imposition of an organizational identity. This not only includes an individual's position in the organization's hierarchy, but also their role in the working of the organization, and the possibilities which are open to them there (Holmer-Nadesan, 1996). As organizations, including the SCC, are historically largely male (both normatively and substantively), it is reasonable to assume that bureaucratic discourses reflect male interests and, that organizational identities are gendered (Ferguson, 1984).

The enclave represents the discursive border placed around women's participation, as women are only encouraged and expected towards activities traditionally associated with normative heterosexual femininity. Discourses of normative femininity construct and reproduce a gendered and sexual division of labour in the SCC, which feminist research argues exists in other organizations (see Savage & Witz, 1992). Women therefore perform supportive labour: labour which has remarkable parallels with work performed in the domestic sphere. Women organize the paperwork, seek out accommodation and nourishment, arrange uniform clothing and repair: labour necessary for the normal routines of the SCC as well as cadet camps and competitions. Women are often placed 'in charge' of the junior cadets (aged 10–12), who as yet do not follow a full RN-based curriculum. Junior cadets are often referred to as the 'babies', and motherly qualities are sought in those who care for them. The constructed nature of women's labour is continuous, demanding, and receives little in the way of reward or acclaim. The gendered division of labour is naturalized to the extent that women are expected to fulfil these secondary and supportive duties.

For women who manage to 'break out' and enter management, the enclave becomes less of a collective boundary, and more of an individual one. Women managers constantly have to negotiate the discursive limitations placed upon their inclusion. They, just like women in other organizations, have to justify and explain their management positions, as men question and undermine women's presence. One female officer described this as "saying why I am here, I should be saying why the hell not." Women in SCC management are also excluded from decision and policy making as their expertise is only drawn upon for 'women's matters'. These range from female uniform to 'girls' problems (from the emotional to the hygienic). Thus the enclave conceptualizes practices of exclusion which operate simultaneously with inclusion. Women are encouraged into the SCC, but then positioned within the organization in ways which effectively exclude them from decision making or from activities which fall beyond the remit of the private sphere.

It is within the context of the enclave that lesbian performances of ambiguity must be viewed. Their performances not only challenge homophobia but also work to resist the positioning of lesbian officers in the enclave. Lesbian officers not only oppose homophobia but the normalcy of institutionalized heterosexuality which dictates normative (hetero)femininity.

NEGOTIATING OUT OF THE ENCLAVE: PERFORMANCES OF AMBIGUITY

Women can reflect upon and challenge the bureaucratic discourses which position them in the enclave (Holmer-Nadesan, 1996). The time of reflection comes when the contradictions and incompleteness of dominant discourses become obvious to individual women. This reflexive awareness creates discursive spaces which, following from Daudi (1986), Holmer-Nadesan terms 'spaces of action'. Spaces of action are effectively women's ability to take advantage of the contradictions of dominant discourses, and to exploit their inability to achieve discursive closure (that is to exhaust meaning). Women can therefore disrupt the imposition of dominant organizational identities through their own reflection and critical practice. Disruption comes in the form of women resisting the imposition of organizational identities and generating their own.

As multiple discourses co-exist within organizations (Leflavie, 1996), women generate counter-identities by drawing upon other discourses to resist or subvert the dominant bureaucratic discourses. However bureaucratic discourses have become dominant by marginalizing other, competing discourses. Those which women draw upon are, therefore, those which are already weakened and marginalized within the organizational site. Women who then resist one gendered identity may well generate another. This was the case for some female officers who resisted some emotional and supportive labours, only to reposition themselves as officers who now make decisions about the same labour. These women may have benefited from equal opportunities policies to promote women to leadership positions, but their organizational identity is still related to the enclave.

The contradictory and limited success of women's resistance could lie in the fact that their resistance is focused upon challenging their engendered organizational identities and not the normative heterosexual relations which shape them. Lesbian officers have more success in exploiting spaces of action as they oppose organizational identities on both gendered (i.e. refusing to perform normative femininity) and sexual grounds (by performing sexual ambiguity). The threshold of the closet, then, enables lesbian officers to resist positioning in the enclave and as a consequence, the gendered division of labour.

Performances of ambiguity disrupt the discursive association between normative sexuality and normative femininity. Lesbian officers refuse to perform consistent pseudo-heterosexuality and thus resist placing themselves in binary relation to men (as 'other'). In doing this they also resist the bureaucratic imposition of the organizational feminine identity. As they constantly resist the 'othering' processes of heterosexuality and femininity, they challenge the notion of their suitability for supportive and secondary labour. I argue that as lesbian officers flirt in their relationship with their defining 'other', they set the whole string of associated binaries into disarray. Performances of ambiguity then disrupt naturalized assumptions of normative femininity, and enable lesbian officers to refuse to be placed in supportive and emotional roles. The threshold allowed lesbian officers to generate an organizational identity which is related to the wider (male) organization as opposed to the enclave. In short, each lesbian officer emphasized her professional relations with cadets and colleagues, as opposed to the emotional.

> I am not like the other women down there. Because they [men] don't know where to put me – gay, straight. It really confuses them. I don't fit. So they sort of treat me like a bloke, but not like a bloke. *Michelle*

> No one sends a crying kid to me. I don't mother them and I get very angry if people assume that that's my role. I am an officer, I hope that I'm fair, strict, but I hope I'm fair. I get respect from the kids. Not because I patch them up when they've scraped their knee. *Fay*

As they resist emotional and supportive labour of the private enclave, they challenge male dominance of the public. These lesbian officers all held positions of middle management and used their ambiguity to transcend 'women's issues' and launch into wider decision making and policy formulating groups.

> They used to try and pigeon hole me, divert me, with little things, stuff they call 'women's issues'. Never quite sure, but its like a whole agenda men don't want to get involved in. In fact, thinking about it, I think they think they're being 'new men' by getting women involved – let them make these decisions. Don't try that with me. I say that's periphery, can be decided later, or say that its not a 'woman's' issue, surely hygiene affects us all, and I force my way into the wider debate, or start off earlier by adding stuff to the agenda. I'm not being forced out. *Fay*

> I'm not the sort that they can just palm things off on. I want it all, I want to know what's happening and who's making what decisions for who. Somehow I can do this, I think because I am seen in my own right, not just good for one thing or another, but a good officer. *Michelle*

Their challenge is enabled by their position on the threshold and their performances of ambiguity. By confusing their location in normative heterosexuality, lesbian officers create spaces of action where they can develop organizational identities which work to further the political effect of the closet. Their resistance of the enclave enables lesbian officers to develop powerful and authoritative voices which challenge homophobia and sexism. Again the SCC's attempt at political correctness is important here, as the lesbian officers not only criticize its limitations, but use it towards their own ends.

> I think that now I'm here, in a job with power, in the man's world if you like, that they've got to listen to me. I speak and they've got to listen because I'm a woman, and they're trying to be pc and that. I've got my feet under their table through their own rules. Some hate it, I think that they don't quite know what to make of it. All of a sudden there's women in this men's club. And its great because they're still trapped into acting with chivalry; even when one of them is patronizing its useful because you can turn it round, show them how simplistic they've made it. What's that expression? Hanging them with their own rope. *Fay*

The success of inconsistent performances that generate ambiguity lies in a continual and conscious performance. Such performances are on-going and tiring.

> Sometimes its easy to give in and just be that mother/daughter person they think they want. D'you know that I presented an idea at a national meeting, and got called 'good girl' by the guy in the chair? Can you believe that? *Fay*

Michelle described her experiences as a 'battle', Debbie as 'always fighting', yet the problems encountered through their resistance were preferable to leaving the SCC, or coming out (which would lead to their resignation). Each of the officers enjoyed the SCC, particularly their relationship with the cadets. They all shared a common vision of leading young people towards responsible citizenship. Their battles against sexism and homophobia harmonized with this vision.

> We keep them off the streets. Some of them can't even read, been given up by teachers and even parents and we give them something. I think we give them self-respect and self-belief, but I'm also aware that we could be churning out hundreds of little bigots. I'm doing something there. *Fay*

CONCLUSION

Lesbian officers use strategies of partial concealment and revelation of their sexuality to resist and avoid their positioning as 'woman' in the private enclave.

They do so by drawing upon alternate discourses which are not yet normalized within the organizational setting. As organizations construct themselves as asexual (Leflavie, 1996) but are, in reality, hotbeds of sexual intent (Hearn et al., 1989), alternative discourses of sexuality disrupt the enclaved border, as they play with the contradictions in dominant organizational discourses. Lesbians who disrupt normative hetero-femininity strike at the heart of the SCC's dilemma over (homo)sexuality. The lack of official policy, pseudo-tolerance and a generally homophobic environment create a space where lesbian officers can manipulate their position from the threshold of the closet.

Lesbian officers flirt from their position on the threshold, performing sexual ambiguity. However a playful anxiety characterizes their experiences, with some lesbian officers tiring of the game, and revealing more than they later think was wise. Damage limitation is then the key, with women playing the 'feminism card'.

> Done it a few times, spoken too passionately then started the next sentence with, 'As a feminist'. I'm not saying this stuff because of my sexuality, but because of my beliefs. *Fay*

This locates their attack upon male attitudes and demeanour from a gendered rather than a sexual base. Although feminism has worked to reject the additive model of oppression, lesbians here, although experiencing oppression at the intersection of misogynist and homophobic discourses, often fight from their (liberated) gendered self, as their lesbian self cannot fully be revealed. This has the effect of empowering themselves as women but renders their lesbianism invisible. Their sexual ambiguity means that their lesbianism cannot be revealed, it can only be playfully suggested, and equally playfully refuted. While their strategies are pivotal to escaping the gendered enclave, each officer was aware of how dependent her performance was upon pseudo-tolerance:

> If someone came up and said, "Are you a lesbian?" that would be the end. I count on them not being able to say that because they are doing their best to be PC. To say this, would mean that they would have to react and this, I think, would just show their hypocrisy. *Fay*

Clarke's PE teachers spoke of the concealment of their lesbian identities as a 'game'. In the SCC, games involve pseudo-tolerance. Lesbians react to it, others work to uphold it without facing its inherent contradictions (homophobia). Lesbian officers are reliant on others to play their game of (pseudo) tolerance, while they, in turn, rely upon lesbian officer to perform their ambiguity, and not their lesbianism. A visible, known and public lesbian would cause a 'reaction' which Fay thinks the SCC wishes to avoid: that is to face up to its contradictions and hypocrisy regarding both sexuality and gender relations. Thus, lesbian officers negotiate the potential and limitations of

pseudo-tolerance. While these enable revelation, they also dictate concealment.

NOTES

1. Fay, Michelle and Debbie agreed to participate as a case study on the
understanding that their identity would be disguised as much as possible. I am using
pseudonyms and I am not providing details of their positions in the SCC, or details of
their life outside. It is enough to say that they are White, able-bodied, with two defining
themselves as middle class and one as working class. Two define their careers as
professional, one is engaged in professional training. Each officer was unknown to the
others.

2. Many non-lesbian/gay members of the SCC found homosexuality and parenthood
incompatible. Conversations I had with SCC members were at the time when the sit-
com *Ellen* was making television history by outing the main character. My fieldnotes
show that in the chats I had with SCC personnel, where we touched upon the *Ellen*
show, many people expressed real surprise when the gay character announced to her
father that her sexuality would not preclude her from having children. Many SCC
people felt that homosexuality and parenthood were incompatible.

3. As a trainee officer I participated in these exercises. They often became an excuse
for people to vent their homophobia. Trainers appeared to sympathize, reiterating that
their 'hands were tied', stating that the SCC would be acting illegally to exclude
members on the grounds of sexuality. I was not able to observe any officer training
sessions during this research but Fay, who had recently officiated on one such board,
shared similar experiences.

4. Fears of child abuse are heightened in uniformed youth organizations. Many of the
male officers who participated in my wider research stress the problems of the 'scout
master' stereotype. Some men discussed how the fear of being labelled in such terms
led them to perform overtly heterosexual masculinity.

5. This recently became headline news when Lord Gilbert the Defence Minister
announced in April 1999 that young gay men and women were welcome to join the
cadet corps. *Pink Paper* issue 572, 26 February 1999.

6. The SCC became gender aware when what had been the separate female
organization was integrated into the male organization. For many staff the 1990s, the
time of the integration, marked the start of a political correctness in the SCC. Senior
management are very keen to avoid charges of sexual discrimination reasoning that
negative publicity would affect the Corps popularity with young women.

7. Cadets often cough out 'lessie' or 'dyke' behind their hands when female officers
who cadets suspect are lesbian, walk past. Many women, regardless of their sexuality,
experience this. Women who are called 'lessies' tend to be young, confident, assertive
and single. The effect of the 'lessie cough' upon closeted lesbian officers depends upon
their position in the power hierarchy and upon their level of confidence and humour at
the time. Fay, Debbie and Michelle reported feeling threatened, worried, annoyed and
amused at different stages of their careers in the SCC. Michelle claims that the lessie
cough has little threat, as most women are targeted in this way. She says, "How can they
know, can they possibly know? They call me a dyke! It's true. They call Sue a dyke –
it's not true – they don't know. Who's the dyke today?"

8. Homophobic remarks and actions do not have an homogenous effect upon their target. Some women are not affected by terms such as 'dyke'; straight women also report experiences of name-calling (lessie) suggesting that this attacks their single heterosexual status as opposed to (any) lesbianism.

REFERENCES

Clarke, G. (1996). Conforming and Contesting with (a) Difference: How Lesbian Students and Teachers Manage their Identities. *International Studies in Sociology of Education*, 6(2), 191–209.

Ferguson, K. E. (1984). *The Feminist Case Against Bureaucracy*. Philadelphia: Temple Press.

Hall, E. (1995). *We Can't Even March Straight: Homosexuality in the British Armed Forces*. London: Vintage.

Hall, M. (1989). Private Experiences in the Public Domain: Lesbians in Organizations. In: J. Hearn, D. Sheppard, P. Tancred-Sheriff & G. Burrell (Eds), *The Sexuality of Organization*. London: Sage.

Hearn, J., Sheppard, D., Tancred-Sheriff, P., & Burrell, G. (1989). *The Sexuality Of Organization*. London: Sage.

Holmer-Nadesan, M. (1996). Organizational Identity and Space of Action. *Organizational Studies*, 17(1), 49–81.

Leflavie, X. (1996). Organizations as Structures of Domination. *Organizational Studies*, 17(1), 23–47.

Leninen, S. (1993). *Gay Cops*. New Jersey: Rutgers University Press.

Savage, M., & Witz, A. (Ed.) (1992). *Gender and Bureaucracy*. Oxford: Blackwell.

SCHOOLGIRL FRICTIONS: YOUNG WOMEN, SEX EDUCATION AND SOCIAL EXPERIENCES

Mary Jane Kehily and Anoop Nayak

INTRODUCTION

This chapter focuses on the experiences of young women in school and their views on sex education. Research in schools suggests that sexuality is present in a variety of exchanges and encounters and can be seen to perform a wide range of social functions (Kehily & Nayak, 1996). This chapter considers the ways in which messages of the official curriculum are mediated by young women and the implications for the production of gendered sexual identities. The research identifies the role of popular culture for young women in the construction of sexual knowledge. The approach outlined here follows a similar path to that forged by the Women, Risk and AIDS Project (WRAP) (Holland et al., 1991a) where a qualitative analysis of the social and sexual experiences of young women is provided. However, unlike the WRAP papers, which consider post-school reflections, our focus is upon young women *at the time of* moving through the education system. The study identifies a range of tensions that exist in the lives of young women in relation to the teaching and learning of issues in relation to sexuality. From the perspective of young women we spoke with, these tensions were exacerbated by reproductive discourses within sex education, cautionary messages from adults and contradictory pressures from young males to have sex and remain virgins. The work illustrates frictions existing at many different levels – *between* the sex education curriculum and

Genders and Sexualities in Educational Ethnography, Volume 3, pages 141–159.
Copyright © 2000 by Elsevier Science Inc.
ISBN: 0-7623-0738-2

the social experiences of young women – and internally *within* each of these
domains. Finally, based on the views expressed by young women, strategies for
the development of good practice in the field of sexuality education are
outlined.

Methodology

This chapter develops out of school-based research which took place over a
period of two years (1991–1992) in two secondary schools in the West
Midlands, U.K. The research project explored issues of sexuality from a range
of perspectives. As researchers we worked with approximately 45 students
altogether, two year 10/11 groups (age 14–16) and a group of sixth formers
(age 16–18). The respondents focused on in this paper are predominantly
white, working class young women, many of them from the school's local
community. These students were selected for us by teachers on the basis of
representing a cross section of the school population, that is young women
from different class backgrounds with different approaches to academic work.
Undoubtedly, the subjectivities of the young women we interviewed and the
context in which the research took place have a bearing on the type of
responses received. In keeping with post-structuralist insights (see Weedon,
1987), we would argue that research encounters produce socially constructed
realities which cannot be regarded as authoritative accounts of the social world.
Rather we would suggest that such accounts can be seen as contextually
specific and socially mediated interactions between respondents and researchers. Our approach to material produced during the research process was to
regard it as discursively produced text which provided us with generative
moments for discussion and analysis. Throughout the research process we
developed a relationship with the young women who spoke to us about their
experiences of sex education and issues of sexuality on many occasions in
different contexts. We would suggest that the accounts of young women we
spoke with offered us glimpses into aspects of adolescent female sexuality from
their perspective which it would be useful for educationalists to consider.
Ethnographic approaches for data gathering were used involving group work
discussion and semi-structured interviews. A tape recorder was used in most
sessions and combined with note taking during and after the research period.
The methodological implications of researching sexuality in schools cannot be
developed within the scope of this chapter. However, we acknowledge the
influence of feminist praxis in this field, particularly its emphasis on reflexivity,
grounded theory and a recognition of the personal (Harding, 1987; Hollway,
1987; Stanley & Wise, 1993).

YOUNG WOMEN AND THE LEARNING OF SEX EDUCATION

Sex education not only brings into focus tensions around gender, but also tensions around generation and the public acknowledgement of adolescent sexuality (Thomson, 1994, p. 40).

In this section we aim to explore the ways young women view the sex education they receive in school. The aim is to identify the main concerns expressed by the young women before considering the relation these have to their daily experiences. The experiences of young women involve the negotiation of sexism and sexist behaviour at many different levels. The chapter argues that sex education for young women does little to challenge existing gender relations and, at times, can be seen to maintain these structures.

Despite Government concerns over social issues such as teenage pregnancy, HIV/AIDS (see, for example, Department of Health [1991] document *Health of the Nation*) and the 'breakdown of the family' many young women experienced sex education as mechanistic and biological.

Libby: We only learnt about frogs and that was about it.

Lucy: Frogs?

Libby: I learnt about frogs in biology, in Middle school . . . I don't remember.

Young women we spoke to felt that the recourse to biology in sex education was of little relevance to their experiences and generally regarded this as inappropriate. The tension between the lived experience of sexual relationships and what is taught in the classroom was made apparent to us in a number of exchanges.

How would your experiences of relationships compare with what you learnt in the classroom?

Amy: It doesn't.

Sally: What they teach us we already know. It's just a waste of time.

Libby: It depends on which teachers you have.

Young women felt that the biological aspects of sexuality overlooked the complex social issues surrounding sexual relationships (Thomson & Scott, 1991). The young women we interviewed were sensitive to these complexities and showed an acute awareness of the absences and fissures within these models.

Melissa: We did sex education in biology but that wasn't very good was it? . . . They didn't go into it with us.

What do you mean?

Melissa: They didn't go into feelings and responsibilities and misunderstandings and confusions and things like that. It was just basically 'take precautions'.

As Melissa identifies, there is more to sex education than reproduction. The biological model appeared abstract and simplistic. The ambiguity of sexual relations where 'misunderstandings' and 'confusions' occur was in stark contrast to the illusory coherence of reproduction referred to in sex education classes. Straightforward reproductive models of sexuality seemed unsophisticated and out of touch with the realities of young women's 'feelings' and 'responsibilities'. Uncontextualized advice such as 'take precautions' makes little attempt to understand pupil cultures and the pressures placed upon young women. The biological model can be seen to have a range of hidden messages that construct the sexual lives of young women. Thomson & Scott (1991) explain:

This style of sex education shapes young women's understandings of what is normal, acceptable and discussible and, while many of them resist and reject these constraints they are nevertheless materially affected by them. What is left out of school sex education is often more significant than the actual content, what is left unsaid can be more powerful than what is spoken (1991, p. 14).

Sex education, then, prescribes forms of appropriate sexual behaviour that has an impact on the lives of young women. Gendered and sexual dimensions are encoded in the language and terms of debate structured through the normalizing discourse of heterosexuality and outlined by teachers.

Melissa: We had this video about sex and that and it made it all very perfect and y'know, all went right the first time and they had this couple making love and there was like music in the background *(all laugh)* and there were a couple of people in our class who have partners and they said, "Oh I don't believe that," and it was very perfect and very straightforward and y'know you love someone, or when you marry – teachers tend to refer to it as *when you marry* you see. We've got a particular biology teacher, and we were doing genetics once and he was going, "When a man and his wife," it wasn't a man and woman it was *a man and his wife*.

In this passage Melissa identifies a *reproductive discourse* used by the teacher and expressed through specific terms and phrases. The film idealizes sexual intercourse in ways that are 'perfect and very straightforward'. The laughter perhaps indicates embarrassment generated by the gap between idealized versions of sex and lived performances. Here, sexually active students appear as a more authoritative source on sexual relations and are able to critique the screen ideal. The bedroom scene with music may seem remote from the realities of young peoples sexual experiences. Hirst (1994) found adolescent sexual activity to be far less glamorous, occurring in outdoor venues such as parks and woods. The discrepancy between pupil cultures and sex education is also underlined by references to marriage. The construction of sexuality based around 'a man and his wife' has specific gender implications that stress the importance of marriage and morality for young women. Heterosexuality and male/female partnership are seen as normative and provides a structure informing the teaching of sex education (Epstein & Johnson, 1994, 1998). Epstein & Johnson (1994) point to the hidden power of heterosexuality and the implications for pedagogic practice:

> Heterosexuality is the silent term – unspoken and unremarked – when sexualities are spoken of. Its invisibility is part of its power. But as the dominant form, preferred in a thousand different ways, it is also the term which must be made visible and challenged if some measure of equality is to be achieved (Epstein & Johnson, 1994, p. 225).

Teacher presumptions of young women as sexually unaware, striving for marriage and implicitly heterosexual, structure gender identities and relations in school. Heterosexist presumptions, moral messages and reproduction were reiterated to young women during the course of sex education programmes.

> *When they talk about sex education, is it in terms of reproduction or pleasure?*
>
> All: Reproduction.
>
> Susan: They just warn you about pregnancy. How it all works on the inside.
>
> Lucy: The sperm and the egg and that.

Female desire, emotions and the pleasure of sex are omitted from the sperm/egg equation (Fine, 1988). As the extract shows, sexuality is presented as potentially threatening to young women who are warned about pregnancy. Similarly, wearing 'provocative' clothing or visiting certain places can be stressed as a sexual hazard for young women (Griffin, 1985).

> Emma: We got shown condoms. In fact, we got shown all sorts of contraceptives and we were told by the head teacher we should never, ever set foot in a pub.

In common with Griffin (1985) and Lees (1986, 1993) our research suggests that young women's' sexual identities in schools are frequently constructed and appraised through discourses of reproduction and reputation.

> Melissa: In junior school we did sex education, in the last year, we had the menstrual cycle, the girls were split from the boys when we did it in junior school. And we had a lady come in to talk to us and then we took a book home and it was mainly about sex. But again very perfect, a man and his wife making love to make a child.

Once again, the idea of sperm/egg or man/woman combining to 'make a child' is invoked through a reproductive discourse. This discourse places constraints on young women by problematizing other kinds of sexual activity. Their perceived roles as future mothers and child rearers is implicit within the reproductive discourse. Indeed, sex education of this type is not apolitical but a means of securing a version of 'the family' and family life that is based upon a matrix of heterosexual, gendered relations:

> 'The family' is a very condensed category, fusing a version of heterosexual relations (love, marriage, and preferably monogamy), conventionally asymmetric relations (separate spheres, the woman as main carer, man as main breadwinner), a stable procreative unit of two opposite-sex biological parents, and a 'moral' environment for children of a strongly normative kind. The *combination* is crucial: the incitement/containment of heterosexual desire, the freezing of gender roles, and the stress on biological and social reproduction – all secured through the same relations (Johnson, 1995, p. 4).

The power dimensions Johnson alludes to are presumed, policed and promoted by the type of sex education many young women receive. However, the need to extend our definitions of sex education, what Redman (1994) more accurately calls 'sexuality education', suggests learning about sexuality occurs across a range of sites. This redefinition attempts to incorporate the social aspects of relationships that young women spoke of: the confusions, ambiguities, fears, desires and a broad repertoire of emotions.

TEACHERS, PARENTS AND PROBLEMS

Although pupils are expected to learn about sex through the curriculum many of the young women we interviewed stressed an accumulation of knowledge which occurred through a variety of informal cultural practices. Within the curriculum, young women felt they were usually treated as 'innocent' and void of knowledge about sexuality. It was felt that much of what is taught in school is repetitive and makes little attempt to engage with prior knowledge pupils may have.

> Libby: The teachers just tell you the basics.
>
> Susan: They just wash over it.
>
> Sally: And Miss Hepworth's too embarrassed to . . .
>
> Libby: You know what they're telling you already.
>
> Amy: They don't tell you anything you don't know, do they?

Once again, young women are critical of sex education and seem to sense an underlying uncertainty on the teacher's behalf. Dealing with 'the basics' or 'washing over' issues was regarded as an evasive strategy employed by teachers.

> Sally: I think the teachers get embarrassed because Mr. Cooper he'd have to like teach the students to put a condom on. I think the teachers just get embarrassed.

Libby: We had this one thing once and the teacher asked us to put, on blank pieces of paper, what your question was and they put them all in a hat and they read them out. Now, that was quite good because anybody was writing anything and the teachers were trying to answer them but that was ages ago.

Susan: Yeah but if the teachers read the question and found them embarrassing, they go, "Oh, we'll come back to that one!"

Sexuality as embarrassing and the effect this has on teacher/pupil relations are featured in the accounts we received. Indeed, the teacher/pupil binary structured the ways in which issues of sexuality emerged within the classroom where, in some instances, sexuality is used by pupils as a vehicle for humour; a point of resistance to disrupt classroom relations and embarrass teachers (Kehily & Nayak, 1996).

Claire: Some of the teachers don't know how to approach it, they're not sure how they should approach it, so it makes it difficult for us to approach them.

In some instances, young women felt they were able to counteract the embarrassment of teachers and find out about sex from their parents. For young women it seemed that it was mothers that tended to provide this information, especially when it came to discussing periods as Amy explains:

Amy: I asked her. I was in junior school and I can remember one of my friends – and she was the first person in the class to start on it – she told me. And I didn't have much of a clue about what she was going on about, so I went home and I asked my Mum about it and she just told me then.

Although some young women had close relationships with their mothers and felt able to talk openly about sexuality this was not always the case. Parents could be seen as a problematic source of information about sex.

Lucy: I reckon it's because if you ask your parents about sex, like, your parents would think you're obviously interested and you want to like . . . That's what I think anyway and I get really embarrassed if I say it to my parents.

So you think they might pass some sort of judgment on you?

Lucy: Yes, they might think that I'm doing it or something.

As Lucy explains, talking about sex to her parents can be embarrassing and arouse suspicions that she is sexually active. When parents did speak with young women about sex it was often in similar terms to teachers with a stress on warning them about the dangers:

Lucy: When I was going out with my boyfriend she goes, "Just remember you're only fifteen. When you're sixteen, a lot of older boys will want to go further."

Right. So that was a bit of cautionary advice?

Lucy: Yes. She said to just beware, just be careful, you know what they're like.

Parental advice, then, is not necessarily dissimilar from that posited by some teachers. The adult sexual warning may also construct female sexuality as passive, or in ways that appear only to respond to male advances. Rachel Thomson describes this 'protective discourse':

> The sex education that girls and boys receive within the home is significantly different. Girls tend to be educated within a protective discourse emphasising the dangers of boys, pregnancy and being 'caught out' while boys are more likely to be teased and encouraged in the development of their sexuality. It is clear that the way young people learn about sex has implications for how they understand themselves as sexual people (1994, p. 56).

The responses of young women also imply that they are sensitive to the social and sexual boundaries constructed by adults.

PEER GROUP PRACTICES AND POPULAR CULTURE

Given the 'warning' approach that many adults provide, young women tended to seek advice about sex from other sources, especially friends.

Sally: I wouldn't be able to ask a teacher. I'd have to ask a friend who'd like experienced it. Because teachers are so difficult to talk to.

Lucy: I usually go to her *(nodding to Sally)* if I've got something wrong. Not that she does that much good – that's nice! No, she does really. I usually go to her.

Researchers have documented the significance of female friendship groups and the role of the 'best friend' for support, particularly in matters considered to be personal and intimate (Griffin, 1984; Hey, 1997; McRobbie, 1978). However, other ways of learning about sexuality existed; young women we spoke with reported using film, books and magazines as cultural resources for learning about sexuality. These resources were a particularly useful means of adding detail and liveliness and thereby supplementing the sex education taught in school.

What's the difference between [what] you learn from magazines and television to what you learn in the classroom?

Libby: There's more information in the magazines and things. More personal . . .

Lucy: They have problem pages and people write in with their problems and you're given like a full explanation and they say, "Well if you want a book, write to this place and ask for so and so book." They give you all this information and it's really good.

For Angela McRobbie (1978) teen magazines play an important part in constructing adolescent femininity through the concept of romance. However, the use of problem pages and teen magazines as a resource for sexual learning may be a means through which teachers can open up discussion in the classroom and attempt to address the social learning that goes on within pupil cultures. The comments we received from young women suggest that information in magazines is regarded as 'more personal' and seemingly more relevant to their needs.

> Amy: When I was at junior school, they never covered like sexually transmitted diseases at all. It was like sex and pregnancy and that was about it. I mean the only way you find out about it is from magazines and stuff.

What became apparent in the research was the ways in which boundaries around sexual knowledge were constructed. Young women seemed aware of these boundaries and the ensuing contradictions of teaching sexuality as if in a vacuum. They were aware of how contemporary soaps such as *Brookside* and *Eastenders* have represented gays and lesbians. They would also frequently refer to sexual episodes in Australian soaps and discuss incidents that occurred. Even so, some of the representations in soaps were treated skeptically by young women who could be critical of romantic ideals.

> Lucy: And then Shane – he's like an eighteen year old boy in *Home and Away* – he's really good looking, has loads of girlfriends and everything and he hasn't slept with anybody. He's like really – over the top.

The responses to sexuality as taught in the classroom and learnt from cultural resources suggests young women do not absorb these ideas uncritically but are able to discern, negotiate and critique information (Barker, 1989; Christian-Smith, 1993; Kehily, 1998; Radway, 1987). A recognition of the ways young women weave together their own cultural patterns of sexual beliefs and behaviour from the social fabric of experience is needed.

> *So who would you talk to?*
>
> All: Friends really, we've all learnt from friends really.
>
> Claire: You just pick it up as you go along, talk to friends, the media, teenage magazines like *Just Seventeen*, talk about relationships . . .

The existing sex education many receive may be, at best, inappropriate or, at worst, damaging with its emphasis on 'moral' values, gendered expectations and structured oppression. Even so, young women are negotiating this complex sexual terrain (Griffin, 1984; Lees, 1993; Skeggs, 1991) and engaging in a process of redefinition within female friendship groups. As we will find in the following sections, current sex education commonly ignores the lived experiences of sexism that young women must work through. It is the multiple tensions produced by sexism that impinge upon the lives of young women.

THE SOCIAL EXPERIENCE OF SEXISM IN THE LIVES OF YOUNG WOMEN

This section will explore the ways young women talk about heterosexual relationships and the sexual beliefs and values they hold. The chapter argues that sexuality education needs to account for the sexist practices and gender inequalities young women encounter (Holland et al., 1991a, b). The previous section considered the ways in which sexuality education is taught, here, we will look at the lived sexual experiences of young women. The exchange below indicates that sexism plays a part in young women's sexual learning. Here, Susan is talking about sex and the surrounding silences.

> Susan: They [*i.e. teachers*] don't tell you if it hurts and bleeds and anything . . .
>
> *But would you be able to ask those questions about does it hurt or anything like that?*
>
> All: No!
>
> Susan: Because of the boys in our class.
>
> Libby: Yes, I think sex education should be separate as well, in some cases.

The research suggests that gender and the power relations inscribed in gendered interactions play an important role in the ways sex education is received and lived out. The impact of these processes upon young women is seen in their relationships, particularly regarding sexual reputations. This section will explore these relations in order to move towards understanding the complex sexual lives of young women against the backdrop of their sex education experiences. We found young women's sexuality could be constrained at a variety of levels where competing demands to have sex/refrain from sex framed their experiences. Here, young women were at the receiving end of the frictions produced by contradictory masculinities.

Relationships and Reputations

> When the dimension of sexuality is included in the study of youth subcultures, girls can be seen to be negotiating a different space, offering a different type of resistance to what can at least in part be viewed as their sexual subordination (McRobbie & Garber, 1982, p. 221).

For young women differing power relations were apparent in the sexual sphere where young men assumed the power to appraise young women through their sexuality. These underlying differences were not recognized by the sex education many young women received; this knowledge was learnt through painful experiences and informal networks of communication. The role of sexuality as a signifier of young women's status was frequently referred to. Young women were aware that they would be judged on their sexual reputations and, unlike males, carried this burden of responsibility. The double standards regarding sexual behaviour was made apparent to us in various exchanges. That they 'cared' deeply about reputations, and young men were generally free of this concern, was a continual source of tension. Here, the young women are speaking about the need to preserve their virginity in the face of male pressure.

Do you think there's pressure for girls to have sex with lads?

Lucy: Yes.

Right.

Libby: It's nothing to them, but it's really important to us.

Lucy: Yes, because once you've lost it, you can't get it back. But when it's a lad, they'll want to just do it straight away, just to like say to their mates, "Oh I've done that. Been there and done that." Just to prove to their friends that they've done things but they don't think about the girls.

Clearly, sexual activity has different meanings for young men and women in schools. The production of these differing meanings is rarely questioned in the classroom. Gendered concerns are expressed in the link between virginity and status when Lucy remarks, "once you've lost it you can't get it back." She could be talking about either her virginity or reputation here, as the two are so interconnected. That these relations occur across a gender binary is revealed by Libby's 'them' and 'us' comment, "It's nothing to them, but it's really important to us." Collective resistance in the face of patriarchal oppression is recognized by some researchers (Jones, 1989; Skeggs, 1991). Whereas young men could experiment sexually, experimentation was not something young women could risk so easily. An 18-year-old male in Willmott's (1971) study describes this uneasy relationship in a seemingly natural way.

You can always get a bit if you want it, with the girls with the big mouths. It gets around that they're that sort of girl. But that sort of thing turns you off after a while – you realise that if you can get it, so can anyone else (quoted in Willmott, 1971, p. 56).

Willmott's respondent illustrates that sexuality can play a far more determining role in the lives of young women where there is continual anxiety that they may become 'the other kinda girl' (Canaan, 1986). For many young women sexual activity with a man was often problematic where 'reputation' is a social determinant on behaviour.

Lucy: I wouldn't do it anyway because I know that you get a reputation here but like – I don't know. It's just that girls get called slags, boys get called studs. Boys get a pat on the back and girls get abuse.

So it's different rules then?

Lucy: Yes. Boys don't get that.

Yes. How do you feel about that then? That there's sort of different rules for boys and for girls?

Libby: I don't know. It just seems that with girls that when someone does – they're called 'slags' and they 'sleep around' and everything. But boys don't do they?

Young women have an understanding of sexuality grounded in experience that can inform the teaching of sex education. The fear that 'you get a reputation' acts as a powerful constraint on the social and sexual practices of young women. These reputations are defined along lines of gendered sexuality where boys are 'studs' and girls are 'slags'. This would suggest that pupil cultures are not homogeneous, but differing peer group expectations exist. Sexuality can either enhance or denigrate pupils' status where "Boys get a pat on the back and girls get abuse." The sexual division of labelling was found in the work of McRobbie and Garber:

> Boys who had, sexually and socially, 'sown their wild oats' could 'turn over a new leaf' and settle down: for girls, the consequences of getting known in the neighbourhood as one of the 'wild oats' to be 'sown' was drastic and irreversible. (McRobbie & Garber, 1982, p. 213)

The research suggests an acute level of awareness existed amongst young women regarding codes of sexual behaviour. Managing sexual contradictions appeared imperative to young women's' reputations. They were also highly insightful as to how these relations are constructed and maintained, something that was not addressed in sex education lessons. We asked about the ways sexual pressure would be applied to young women and they identified masculinity as an important social factor.

So who would the pressure [to have sex] come from then?

Lucy: The lads. It's mainly from the lad's mates actually, I reckon.

Right. So you think the lad's actually under pressure himself?

Lucy: Yes, the lad's under pressure himself to do it. I've heard so many stories about that.

The response suggests a need to develop more research in the areas of pupil cultures, masculinity and sexuality (Mac an Ghaill, 1994). The passage implies that biological notions of male sexual drives do not account for the complexity of sexual relations within peer group cultures. Formations of masculinity and the ways in which gender positions are socially negotiated highlight the inadequacy of viewing sexuality in these reductive ways. The social expectations placed on young men to be seen to perform, complicates their relations with women. The young women we spoke with indicate an awareness of this and a realization of the differing burdens and consequences for males and females. Although young men were also subject to gaining sexual reputations the meanings were structured through gendered power relations as Wolpe explains:

> Like some girls, some boys also developed reputations, but the major differences was that the boys stood to gain from having a reputation of being sexually active, while . . . it could be harmful to girls. Boys' sexuality is equated with masculinity whereas girls' femininity resides in their outward appearance and behaviour which specifically excludes sexual experience thus reflecting the double standards which have operated in society for well over a hundred years (Wolpe, 1988, p. 167).

The influence of the male peer group upon the sexual lives of young women was remarked on in several instances.

> Lucy: Oh, I was walking through the park the other day, one of Michael's friends was going, "Oh when are you going to do so and so *(i.e. have sex)* with Michael?"
> And I said, "I'm not. I'm not ready." And he goes,
> "Oh just do it. It's only like, only . . ." *(inaudible)*
> "No, I'm not ready."
> So I think like they expect it. So that did my head in a bit.

It is through these encounters that young women are assessed and tested. The psychological pressures they are under is revealed when Lucy adds that it "did my head in a bit." There is an expectation that sexual intercourse is more acceptable for young women in longer term relationships (Lees, 1993), though at what point this is appropriate is unclear. The public and private tensions regarding sexual behaviour were identified by other young women.

> Sally: Some boys like, go back to the lads and say, 'Oh, I did this with her last night' which is wrong. They should keep it between them and the girlfriend.

Given the power and pervasiveness of sexual rumouring and the effects this can have, it is not surprising that many young women monitor their behaviour so carefully. The significance that pupil cultures play in actively constructing sexual relations has led Carol Lee to emphasize the role of the 'nuances, gossip and rumours that are doing the rounds' (1986, p. 23). The policing of young women's sexuality and the relationship with patriarchy are highly significant as Sue Lees explains.

> Girls walk a narrow line: they must not be seen as too tight, nor as too loose. Girls are preoccupied in their talk with sexuality, and in particular with the injustice of the way in which they are treated by boys. Defining girls in terms of their sexuality rather than their attributes and potentialities is a crucial mechanism for ensuring their subordination to boys (Lees, 1993, p. 29).

In the earlier parts of the chapter we saw how sex education may discourage the exploration of adolescent female sexuality where cautionary messages find resonance with parental warnings. Although the subjects of the research seemed to indicate that most young men were 'only after one thing' other ways of controlling young women's sexuality existed. 'Moral' policing, especially by young men, has an influence on the lives of young women and is usually enacted through the virgin/whore dichotomy Griffin identifies:

> The virgin/whore dichotomy has been, and remains, of central relevance to patriarchal power. It creates a double standard which is currently experienced by many young women through the need to guard their sexual reputations, and cope with male demands for sex (Griffin, 1982, p. 11).

For Griffin, then, the sexual sphere is an already determined landscape where young women are ideologically and structurally subordinated. In the following discussion young women are talking about the attitudes of certain males to female sexuality.

> Libby: I can remember Simon Irons saying that Dave Smithers thinks it's immoral to do things like that (have sex).
>
> Sally: Oh yes, I can remember him saying that.
>
> Amy: He's just the same!
>
> Sally: Yes, but I couldn't see Dave Smithers doing it though!
>
> Susan: No, because I can remember Simon Irons saying in English that Dave Smithers didn't want to have sex with Sharon because she was under age and he didn't want to pressure her or something. That's what he's supposed to have said.
>
> Susan: And he gives them all the rights on smoking and drinking and everything!

Sexually demanding masculinities are, then, only one level of sexism young women must negotiate. It seemed young men were influential in setting the sexual agenda, where women were expected to respond to these demands. Female desire was often silenced within relationships due to their concerns with issues of reputation. In the above example men may act as 'moral' guardians defending women's (and subsequently their own) sexual status. The assertion of a 'moral' masculinity that seeks to control and protect the sexuality of young women was a feature of the research. However, it was not only boyfriends that could enact this, but male peers and family members too. Here, Susan is speaking about an episode in a soap opera and the role of fathers as 'moral' guardians.

> Susan: In *Home and Away*, if I just went out for the day and then 'phoned up to tell my dad that I was in a hotel room with somebody, I think he'd come down and kill me!

It did appear that young women could internalize these masculine morals and were perhaps less inclined to resist these than the strong sexual masculine desires for sex. The self-policing anxieties of negotiating the virgin/whore dichotomy were seen in examples where female resistance to heterosexual intercourse was frequently not very empowering and often worked alongside a masculine morality. Perhaps, this is why Griffin claims "The dichotomy between Good and Bad women is not a straightforward ideological division which can be negotiated. It is a profound *contradiction* in which young women always lose, whatever they do" (1982, p. 11). Here, we get an insight into how these subtle sexual relations work.

> Lucy: A boy said to me the other day – this kid I know who's left the school two years ago or something like that – he said,
> "I hope you don't start sleeping around because all the boys will be wanting like to be around you for ages and after about a few months, like when you've slept with all the people, they're not gonna want to know you. They're just – they'll just move on to the next best thing, if you see what I mean."
> So I said, "I'm not going to."
>
> Susan: Who was that?
>
> Lucy: Lesley Adams *(snickers)*. He's a boy called *Lesley*, but he was just saying that . . . I took that as a good piece of advice.

In some cases it appears young women and men may enter into a shared discourse of control and desire where there is agreement on appropriate sexual behaviour. Rather than questioning the masculine policing of sexuality, young women's identities were invariably shaped by these processes. Lucy then takes the young man's comments 'as a good piece of advice'. Despite the pressures of male peer groups it was not impossible for young men to develop a less performative sexuality, although this was generally regarded as exceptional. One male, Nathan Briggs was described to us as 'different' because 'he isn't big headed' and 'you can talk to him'. Reading this extract we get a sense of the social pressures of male peer groups and the direct ways it impinges on the lives of young women.

> Lucy: This girl that he sees – he's only been seeing her for about two weeks, a week, hasn't he? – anyway, yesterday all his mates were saying, "Oh take her in the bushes! Take her in the bushes!"
> What did he do? He just walked off didn't he!
> He just like ignored them. He just like kissed her.
>
> Susan: Aah, that's really nice.
>
> Lucy: Now, if that was anyone else, they'd have gone, "Yeah, come in the bushes, come in the bushes."

The way the young women talk about Nathan Briggs suggests his behaviour is exceptional. We also get an insight into the sexual cultures of adolescent males where sexual performances and calls to "take her in the bushes" are voiced. In his study on male sex talk, Wood (1984) found that sexual abuse was not only about defining women, but also a way of constructing masculine identities as such verbal interactions seek to "sort out the whole female gender and, perhaps, *the emotions of the speaker*" [original emphasis]. (1984, p. 60).

We have already seen how young women may be aware of the reasons why males performed sexual games and talk brazenly about their exploits. Even so, this did not necessarily make it any easier to challenge sexist actions. Being able to spot the sexual double standard and simultaneously challenge it was not simple (Cowie & Lees, 1987; Lees, 1993). Young people may also have different attitudes to sex education linked to the social construction of gender.

Do you think boys are as informed as girls are around sex education?

Sally: Boys think they know it all. They think they know everything.

And in your experience?

Lucy: Boys are more open about relationships and sex and that lot. I think girls try and keep it behind closed doors because it's like their personal life but boys try to show off but I know that I'd try to keep it to myself if I was doing what most of the boys in our year are doing.

It is apparent that young men are encouraged to speak about sexual relations (Haywood, 1996; Willis, 1977; Willmott, 1971; Wood, 1984) and 'show off' often at the expense of women. The gendered dynamics of acceptable sexual practices allow young men to appear 'more open' about sex and relationships, whereas young women are discreetly handling a range of contradictions. The fear of getting a reputation necessitates these responses. That divisions along lines of gender, 'race', class, sexuality and a plethora of complexities exist amongst pupils is itself highly significant. These differences need to be addressed within sex education or inequalities will continue to occur within and beyond the classroom.

CONCLUSIONS: STRATEGIES FOR THE DEVELOPMENT OF SEXUALITY EDUCATION

Qualitative research that has explored the ways in which young men and young women understand sex and sexuality and the way in which they related to dominant ideals of masculinity and femininity provides an alternative empirical basis from which educational initiatives can develop (Thomson 1994, p. 54).

The experiences of the young women we spoke to suggest they have much to contribute to the development of sex education programmes in schools. The following points emerge from their accounts:

- The reproductive discourse for sex education is problematic. It bears little relevance to the complex social world of young women. It can reinforce their subordination through the generation-based emphasis on morality, marriage and motherhood and is thus inherently heterosexist.
- Definitions of sex education need broadening to account for the 'confusions', 'misunderstandings' and range of emotional struggles that are part of sexual relations.
- A student-centred approach is helpful here to challenge the pervasive teacher/ pupil binary and 'personalize' practice. Simplistic adult 'warnings' are unhelpful and position young women as requiring protection.
- Popular culture can be seen as a resource for sexual learning and can be utilized by teachers for discussion. Young people do not absorb ideas passively but are active in constructing beliefs from a range of sources.
- Pupils are not 'innocent', some are sexually active and all are aware of the significance of sexual cultures to their identities. This means young women are frequently at the receiving end of the contradictions of masculinity.
- Inequality exists producing a multitude of frictions in the daily lives of young women. These inequalities must be challenged and sometimes single-sex classes may be necessary to address certain issues. Teachers need to be alert to the frequently misogynist and homophobic structure of certain pupil cultures.

The points raised are not intended as a rigid blueprint for teaching sex education. However, they do help establish a framework aimed at meeting the needs of young people where certain approaches are especially inappropriate for young women. Since these strategies have grown out of the responses of young women it is hoped that the issues raised will be of use to teachers, educationalists, feminist practitioners and young people in need of a more socially specific sexuality education.

REFERENCES

Barker, M. (1989). *Comics, Ideology, Power and the Critics*. Manchester: Manchester University Press.

Canaan, J. E. (1986). Why a 'Slut' is a 'Slut': Cautionary Tales of Middle Class Teenage Girls' Morality, In: H. Varanne (Ed.), *Symbolizing America*. Lincoln: University of Nebraska Press.

Christian-Smith, L. (Ed.) (1993). *Texts of Desire, Essays on Fiction, Femininity and Schooling*. Lewes: Falmer Press.

Cowie, C., & Lees, S. (1987). Slags or Drags. In: Feminist Review (Eds), *Sexuality: A Reader*. London: Virago.

Department of Health (1991). *Health of the Nation*. London: HMSO.

Epstein, D., & Johnson, R. (1994). On the Straight and the Narrow: The Heterosexual Presumption, Homophobias and Schools. In: D. Epstein (Ed.), *Challenging Gay and Lesbian Inequalities in Education*. Milton Keynes: Open University Press.
Epstein, D., & Johnson, R. (1998). *Schooling Sexualities*. Buckingham: Open University Press.
Feminist Review (1987). *Sexuality: A Reader*. London: Virago.
Fine, M. (1988). Sexuality, Schooling and Adolescent Females: The Missing Discourse of Desire. *Harvard Educational Review, 58*(1), 29–53.
Griffin, C. (1982). The Good, the Bad and the Ugly: Images of Young Women in the Labour Market. *CCCS Stencilled Paper*. University of Birmingham, no. 70.
Griffin, G. (1985). *Typical Girls?* London: Routledge and Kegan Paul.
Harding, S. (Ed.) (1987). *Feminism and Methodology*. Bloomington: Indiana University Press.
Haywood, C. (1996). Out of the Curriculum: Sex Talking, Talking Sex. *Curriculum Studies, 4*(2), 229–249.
Hey, V. (1997). *The Company She Keeps: An Ethnography of Girls' Friendships*. Buckingham: Open University Press.
Hirst, J. (1994). *Not in Front of the Grown-Ups*. Health Centre Report no. 6. Sheffield Hallam University: Pavic Publications.
Holland, J., Ramazanoglu, C., Scott, S., Sharpe, S., & Thomson, R. (1991a). *'Don't Die of Ignorance: I Nearly Died of Embarrassment' – Condoms in Context*. London: Tufnell Press.
Holland, J., Ramazanoglu, C., Scott, S., Sharpe, S., & Thomson, R. (1991b). *Pressure, Resistance, Empowerment: Young women and the Negotiation of Safe Sex*. London: Tufnell Press.
Hollway. W. (1987). *Subjectivity and Method in Psychology*. London: Sage.
Johnson, R. (1995). Contested Borders: Contingent Lives. In: D. L. Steinberg, D. Epstein & R. Johnson (Eds), *Border Patrols*. London: Cassell.
Jones, C. (1989). Sexual Tyranny: Male Violence in a Mixed Secondary School. In: G. Weiner (Ed.), *Just a Bunch of Girls*. Milton Keynes: Open University Press.
Kehily, M. J., & Nayak, A. (1996). The Christmas Kiss: Sexuality, Story-telling and Schooling. *Curriculum Studies, 4*(2), 211–227.
Kehily, M. J. (1998). Learning about Sex and Doing Gender. In: G. Walford and A. Massey (Ed.), *Studies in Educational Ethnography, 1*. Stamford: JAI Press.
Lee, C. (1986). *The Ostrich Position: Sex Schooling and Mystification*. London: Unwin.
Lees, S. (1986). *Losing Out*. London: Hutchinson.
Lees, S. (1993). *Sugar and Spice*. London: Penguin.
Mac an Ghaill, M. (1994). *The Making of Men*. Milton Keynes: Open University Press.
McRobbie, A. (1978). Working Class Girls and the Culture of Femininity. In: Centre for Contemporary Cultural Studies (Eds), *Women Take Issue*. London: Hutchinson.
McRobbie, A. (1981). Just Like a Jackie Story. In: A. McRobbie & T. McCabey (Ed.), *Feminism for Girls: An Adventure Story*. London: Routledge and Kegan Paul.
McRobbie, A., & Garber, G. (1982). Girls and Subcultures. In: S. Hall & T. Jefferson (Ed.), *Resistance through Rituals: Youth Subcultures in Post War Britain*. London: Hutchinson.
McRobbie, A., & Nava, M. (Eds) (1984). *Gender and Generation*. Houndmills: Macmillan Education.
Radway, J. (1987). *Reading the Romance: Women, Patriarchy and Popular Literature*. London: Verso.
Redman, P. (1994). Shifting Ground: Rethinking Sexuality Education. In: D. Epstein (Ed.), *Challenging Lesbian and Gay Inequalities in Education*. Milton Keynes: Open University Press.

Stanley, L., & Wise, S. (1993). *Breaking Out Again*. London: Routledge.

Skeggs, B. (1991). Challenging Masculinity and Using Sexuality. *British Journal of Sociology of Education, 11*(4).

Thomson, R. (1994). Moral Rhetoric and Public Health Pragmatism: The Recent Politics of Sex Education. *Feminist Review, 48*, Autumn.

Thomson, R., & Scott, S. (1991). *Learning about Sex: Young People and the Social Construction of Sexual Identity*. London: Tufnell Press.

Walkerdine, V. (1990). *Schoolgirl Fictions*. London: Verso.

Weedon, C. (1987). *Feminist Practice and Poststructuralist Theory*. Oxford: Basil Blackwell.

Willis, P. (1977). *Learning to Labour, How Working Class Kids get Working Class Jobs*. Farnborough: Saxon House.

Willmott, P. (1971). *Adolescent Boys of East London*. Middlesex: Penguin.

Wolpe, A. M. (1988). *Within School Walls: The Role of Discipline, Sexuality and the Curriculum*. London: Routledge.

Wood, J. (1984). Groping towards Sexism: Boys' Sex Talk. In: A. McRobbie and M. Garber (Eds), *Gender and Generation*, London: Macmillan.

TROUBLING THE AUTO/BIOGRAPHY OF THE QUESTIONS: RE/THINKING RAPPORT AND THE POLITICS OF SOCIAL CLASS IN FEMINIST PARTICIPANT OBSERVATION

Valerie Hey

INTRODUCTION: POSTMODERN SOCIOLOGY – ONLY FOR THE SMART AND SUICIDAL?

I think there is an unhealthy end-of-century pessimism in much contemporary social critique. This is linked to, and fuelled by, a tendency I have noted elsewhere (Hey, 1999a) for political disengagement on the grounds that postmodernism mandates an endless deferment of 'what's to be done'. However, to argue against postmodernism risks capture by either of its implicit 'serviceable others' (Morrison, 1992) namely, the voice position or, even more problematically, complicity with the same conditions of power one is struggling to theorize and challenge. What I wish to explore here is the central paradox that research is always at some level about seeking and in part claiming an understanding of the 'other'. Moreover, our (provisional) interpretations arise in participative qualitative enquiries partly out of claims about 'successfully' negotiating difference. As a result, investments in rapport are steeped in 'contaminated' humanism with its legacy of authenticity and essentialism. Yet postmodernism puts this 'modernist' project under severe

Genders and Sexualities in Educational Ethnography, Volume 3, pages 161–183.
Copyright © 2000 by Elsevier Science Inc.
All rights of reproduction in any form reserved.
ISBN: 0-7623-0738-2

theoretic pressure (Scheurich, 1997) because it is precisely Western discursive forms of knowing that are discredited as 'imperial sameness once again' (Foucault, 1972). But if we cannot 'do' any of our understanding and interpretation on this basis how can we continue to undertake qualitative work, much less write about research findings? These are important concerns, especially if we are interested in participative methods. Some propose that research is now an impossibility, a position endorsed by an eminent male social critic at a recent educational research conference. Others suggest experimenting with a new representational politics of poly-vocality (Lather & Smithies, 1998). My alternative is to try to rehabilitate rapport.

CONVENTIONAL UNDERSTANDINGS OF RAPPORT: ERASURE OF DIFFERENCE?

Studies premised on sustained fieldwork (like participant observation) work in sensitive areas and research which explores aspects of the previously little known could hardly be undertaken *without* rapport. But what is meant by the term rapport? It is generally understood as the response to the researcher capacity for interpersonal dexterity, leading to a measurable outcome – a good interview. Women are said to be good at 'it' (Oakley, 1981), the consequence of their domestic socialization into interpersonal competencies such as emotional flexibility, sensitivity and good humour. Its synonyms are: affinity, camaraderie, compatibility, harmony and understanding. A researcher 'good at rapport' occupies the position of a hybrid deity – between the vocation of priest and earth mother.

In this model, rapport is both cause and consequence of warm mutual interactions. These arise as the result of being a good questioner, observer and listener. These capacities allow the 'other' to display her (self) understandings so as to minimise self-censure. The interviewee is encouraged to be unguarded and engaged. As researchers our intent is on reducing the distance between ourselves and the subject of our research. Yet there are clearly problems with this notion. Viewed as seduction rapport/empathy can also be seen as one of the best tricks of the exploiter's trade (Finch, 1984; Stacey, 1988). Moreover, in conditions of contemporary 'fast feminism' (Kenway and Langmead forthcoming) a particularly problematic form of alienated labour – allowing star academic performers to strut their clever analytical stuff over others dead and vampirised bodies. If this is *all* there is to rapport and research then we might well understand the exclusion zone some academic feminists (such as Lather, 1998) seek to construct around it. But rapport is *not* the totalizing concept presumed by difference postmodernism. In the field it does *not* lead to the automatic re-inscription of power relations – it can lead instead to their temporary destabilization and reconfiguration.

POSTSTRUCTURALISM AND THE CASE FOR RECONCEPTUALIZING RAPPORT

At issue here is a mistaken conceptualisation of social inter-relations as either/ or 'same' or 'difference'. In reconsidering my own work (Hey, 1997a) I highlight another possibility derived from a post-structuralist understanding of social relations. This is a position which views social relations as co-constructions established in conditions of social power through a matrix of binary relations. As numerous writers have argued (Davies, 1993), a concept such as femininity does not make sense unless set against its binary 'opposite' notion of masculinity, nor does the idea of whiteness gain power until set up as the 'superior' to its 'inferior other' of blackness (Hall, 1992). A central point of the ensuing argument rests on seeing 'rapport' as a momentarily achieved (fragile) form of inter-subjective synergy that is produced with and against the force of the binary norms that marshal us into our respective places. Rapport can be thought of as a form of inter-personal hegemony. This is a position which disputes that empathy/rapport construct the cancellation of all differ-ence(s) – a concern implied by strong post-modernist difference positions. Rapport is always a process and thus always on the edge of destabilizing. I want in the following discussion to suggest why.

I take the construction of rapport as a material practice in two ways: first, as a socio-cultural basis for developing and sustaining inter-personal meaning-making in the course of field work and second, as indexical of the power and purchase of prior social experience. Looked at from these perspectives rapport is never fully achieved, stable or without fracture. As with human relations more generally, inter-personal rapport in fieldwork is wrested from the same complex relay of inter-personal/intra-psychic relations of sameness and 'difference' through which discourses of power/powerlessness are deployed.

I assert this from my own practice since I have never felt the force field of research relations as singular monolithic states of rapport/'difference'. My experience suggests that research exchanges are messier, more ambiguous encounters – processes of connection, disconnection, break, rupture and re-connection spilling outside of simple binary modes. I discuss the material practice of rapport next.

KNOWING ME, KNOWING YOU. DREAMS OF ESCAPE: PART ONE

I have previously acknowledged in my recent book about girls' cultures of friendship (Hey, 1997a) the power of experience, including the recursive production of shifting forms of split affinities/allegiances in thinking about and representing 'Carol':

> Somewhat more complexly and unconsciously, my analysis also reflects my own memories of being a different sort of working-class schoolgirl. Carol's story therefore is not only built through my recontextualisation of aspects of her own representations of herself, it is also mandated by inter-subjective forces – of memories of growing up with and against girls like Carol (see Chapter 1). In fact it was only in re-reading the original manuscript that I became aware that I had chosen the name of one of my former best-friends as her pseudo-name!

> . . .In particular, the fact that I am telling Carol's story rather than she telling mine, dramatises the prevailing class and age relations between us – which no amount of empathy or feminist commitment or sharing of the project can dissolve. The ability to tell another's story is a concrete social practice of power and it crystallises for me the distance travelled out of my culture of origin. I both grew up with 'Carols' and grew apart from and against them. Carol self-consciously mimicked the differences between us in the last days in the field at Eastford – she followed me around pretending to write her observations into a book! The 'text' of Carol emerges out of the uneven interplay of all of these factors (Hey, 1997a, p. 89).

Whilst the conscious and unconscious (intuitive?) dynamic of interpretation was present in respect of all the girls in the study, it was felt most intensely by me around the politics of analysing the data generated by and about 'Carol'. This was primarily because of the closeness of our relationship. This 'intimacy' promoted in my imagination a proxy for 'who I once was'. This implicated the production of moments of recognition. 'Understanding Carol' was produced by complex exchanges between the psychic as well as the rational. But what do these multiple layerings of knowledge do to how we can conceptualize field relations and data analysis?

Implicit in negotiating and sustaining 'access' are fine-tuned readings of what is or is not permissible. Certainly, the fact that I had at some level also 'shared' aspects of working-class girls' biography[1] as well as certain middle-class girls' investments in education provided useful (and at times essential) cultural and inter-personal resources as did my prior experience as a secondary-school teacher. My own experience also impacted on the interpretations of girls' practices. Understandings were reliant on the following resources: observations in the field, conversations and data that 'Carol' and other girls shared with me; my grasp of the theoretical literature and other empirical studies and much less systematically influenced by 'memory work' (Crawford et al., 1992). Neither could I avoid using aspects of my own present experience as a mother of two highly social daughters!

In revisiting these complexes of knowledge, I want to focus on my reading of one particular girl. I want to extend my earlier discussion of these relations/ splits as a resource for recognising what ground of same/difference was worked between us in the course of fieldwork as well as within the politics of representation. Invariably, this means risking further self-exposure which, as Mary Kehily elegantly poses, is never an innocent and always a performative (possibly a narcissistic) activity (Kehily, 1995). I propose this exploration in order to ground rapport as the work of self-reflexive subject(s). It is neither magic nor accident but a praxis that can assist in producing (imperfect) self/ other understandings. This position is distinct from identity-essentialism in which the self/other or researcher/researched is conceived as impermeable binaries and from deconstructivist end-games of 'difference', since neither position leaves space for recognitions of what Phil Cohen has described as the 'other within' (Cohen, 1997). When I returned to study girls who were living similar backgrounds to the one I had 'escaped' (Hey, 1997b) I felt it as an imaginative re-entry to a strangely familiar territory, despite it being experienced 30 years apart and at different ends of the country.

WORKING CLASS GIRLS' (DIFFERENT) DREAMS OF ESCAPE

Anne Whitehead found in her study of rural Herefordshire that dreams of escape including 'running away' represented the main female fantasy (Whitehead, 1976; cf. Sherratt, 1983). Here are two accounts of this theme from my study. One is authored by me the other by 'Carol':

> Carol reports that her mother keeps a very tight supervisory rein on her whereabouts and yet she also tells me that she is allowed out 4/5 days a week! She additionally says that she had escaped out of the house for a 2.30 a.m. rendezvous with Ivan (One of her boyfriends) near the river. She had apparently used the drainpipe to effect her freedom, like all good school-girl heroines! (Field notes 7).

Carol also later wrote this note to her then best friend Liz:

> Dear Liz,
> I don't give a fuck about home, I'm leaving, don't forget when you're 16 you're leaving too.
> Are you coming?
> Love Carol.
> Sorry about the red ink. It's the only one I could find.

Uncovering these motifs of escape made me think about my own 'fantasy' life as a 14 year-old working-class schoolgirl. It was certainly stimulated by reading similar magazines, but my re-working of the narrative took a different tack. It was our very different constructions of imagined future that is particularly salient.

CLASSIFYING DESIRES; POSITIONING HETEROSEXUALITY, DOMESTICITY AND SCHOOLING

Carol's domestic workload was premised on the core assumption that she would 'help'. Home was not a work-free zone, as it was for so many of the middle-class girls I interviewed (and in part envied and resented). Yet, if she (like me at her age) was immersed in domestic labour, she also appeared to be more transgressive than I ever dared. She was feistier and appeared to relish 'behaving hard' (Hey, 1997a) – a position which gave her apparent power as she struggled and battled for autonomy. This was waged over control of her household labour and her sexuality. Her assertion scared me. It pulled me back to memories of similar girls I had feared. I can still see some of them now. These were girls who were sexually provocative, who always seemed much more 'advanced' in their bodies and habits, who looked out to the world beyond home and school. These were girls with 'figures' (who made me feel a thin sexless androgynous beanpole). These girls seemed *au fait* with make-up, fashion and 'dating' and secure in the esteem gained as the objects of (heterosexual) desire in both secondary school and in the follow-on further education college. I, in contrast, was the serious studious 'speccy-four eyes' intent on education as the only route through to a 'respectable' (Skeggs, 1997) girlhood with its boundaries of unfathomable chastity/celibacy.

In contrast, Carol positioned herself completely differently. She represented herself as the heroine of an altogether 'sexier' aspiration – the fictional/fantasy – heterosexual romance. As other feminist research has suggested romance is still the only ideological category that allows girls (working class ones in particular) access to 'sex' if it is for 'for love'. I have also argued that romance is the only credible strategy that appears to give white working-class girls claims to men's money via the rituals of courtship (Hey, 1982). This is Carol's first entry in the research diary she kept for me (Carol's Research Diary):

> This morning my mum woke me up. She asked me to look after the baby while she went to work so I said "Yes alright then." Afternoon struck so I took the baby for a walk down the High Road. I was listening to my head phones when Peter passed me. I shouted back to him and he stopped and walked back a few paces then said "Hi".

"Hi". I replied.

"What have you been doing with yourself?" said Peter.

"Not much haven't seen you lately and where's Fella?" (Peter's dog).

"Well I moved to Centre Green and in my flat you're not allowed puppies or a cat."

Oh Fella was such a nice dog as well.

So he said all of a sudden "When are we going to make love?" I was surprised so I said "Don't know."

"Meet me over Garden House by the cafe" Peter said.

"Ok" (I) said.

I went over there at 1.30 p.m. and waited. He didn't turn up so I started to walk home. As I had just come up the subway there Peter stood talking to his mate. So I stood there waiting finally he came up to me and said

"Sorry I didn't meet you but I went everywhere to find my trainers."

"Yeh that's alright."

"Forgiven?"

"Of course Peter you know I have to forgive you, don't you, because you are sweet!"

"Walk me to the other end of the subway?"

"Yeah, why not."

I walked Peter up there then he kissed me goodbye.

I said to him "Will you be down the Pond Cafe?" so he said

"Yeah I still hang around there a lot"

So I said "Bye Peter" and went.

Carol's style captures the conventions of romance, with its theme of the saturation of everyday life with sexuality. The text is accomplished through the discursive erasure of 'reality' by 'romance'. The 'Peter Story' starts with her in the role as her mother's domestic – shopping and taking her baby sibling for a walk but meeting Peter works to transform the drudgery of her enforced child care work into the 'glamour' of her chance encounter. Cinderella motifs connect to popular consciousness precisely because they connect to working-class girls' immersion in household labour. Carol's diary entries usually begin with jobs or tasks she has been set by others; her mother; her younger brothers etc. – but then domestic 'skivvying' invariably lead to erotic encounters with a variety of boyfriends. Walking her dog or walking the baby are particularly helpful narrative and social devices for meeting men.

Whilst the above text converts the banality of domestic life into a 'brief encounter', tracing a journey from boring chores to high drama – her other diary which I decided was too personal to include in the book. Carol's Personal Diary tells a counter-romance story from the opposite direction. In sum, it traces the reverse journey from romance to domesticity and tracks a story of tenderly represented but still 'dangerous liaisons' – tender because they are stories of care and passion but dangerous because the passions they represent threaten their subject Carol with losing control over her body. The discourse of romance in which she placed herself as both a desirable and desiring subject 'turning men on', 'feeling horny' gives way under her recognition that premeditated 'protected' sex is not consonant with spontaneous 'romance' nor traditional working-class mores about female sexuality, viz. premeditation = slag. Carol brings the text to a close with worries about the fear of being pregnant and ends the diary by planning the name of the desired/'dreaded' child (see and cf. Steedman, 1982). "If it's a boy it'll be called Adrian Lee after the daddy. Girl – Lily Audra after my mum!" Here the aspirational themes of desire and power of heterosexual romance as a 'proper young women' is literally as well as materially grounded (as well as being ground down) by working-class girls' lack of sexual and social autonomy.

It was painful to recognize in 'Carol' the cost, not only of her own material 'choices' but also to confront the psychic costs of undertaking my own very different 'escape', which was necessarily sexless, sanitary and hygienic! When I was about Carol's age (14) I too produced an inspired 'dream scape', but my 'escape' was imagined as neither drainpipe nor dramatic. On the contrary, my account was public – commissioned by a school 'housecraft' task. The result was a highly invested lovingly detailed narrative of me as the housewife superstar in my 'ideal home'. This was an imagined future conjured of gadgets and cleanliness. If it was escape – it was an escape from one household where I actually did the majority of household work (ironing, cleaning, the laundry to the laundrette, making meals, washing up, child care etc.) to a much more affluent middle class world of semi-detached homes, appliances and dream kitchens! In this fantasy I had shifted myself from being my parent's household labourer to being my own! I do not recall anyone else in this narrative, just me and the pleasurable absorption in singular tidiness. As fantasies go it is odd, both asexual and antiseptic, perhaps because in 'respectable' working-class northern cultures 'nice' daughters did not 'let men near enough'!

So when my friends were getting caught by their mothers with secret diaries with their less than subtle sexual codings, I was on my alternative quest – for cleanliness and 'respectability', an identity that as Beverley Skeggs has argued (Skeggs, 1997) demanded precautionary self-monitoring for the production of sexual and social modesty. My private diaries at the time were dismal affairs, full of earnest mentions of homework and dentist appointments. Bubbling along within, though, were clearly ambitions for 'something else', yet this desire had to be mindful of injunctions of 'not getting above my station'; 'drawing attention to myself'; or 'airs and graces' or 'lasting and pasting' (Walkerdine, 1985). What this implied was an identity that demanded diffidence in matters both sexual and educational – effectively androgynous or asexual – a sort of social aspirational prophylactic – a cordon sanitaire. My mother's injunctions against 'dirty necks' and the absolute imperative of having 'clean clothes' suddenly took on a new urgency. In fact scrubbing clean was the metaphor of the body regime that I most remember. The emotion of shame is implicated here. I tactfully selected *The Good Housekeeping Guide* for my prize in English in 1963! Moreover, I also studied Domestic Science at A Level which tells you all you need to know about translating opportunity structures for accommodating white working-class girls in the further education colleges of the 1960s!

My own dreams of escape thus reflect another version of the story about the romance of an ill-defined 'glamorous' life beyond my council estate world. Going back to school as a researcher was thus a necessary confrontation with the girl I once was before 'passing' (Kuhn, 1995) before the wanting and desiring (Walkerdine, 1985). My ethnography could not help but provoke memories of my own girlhood and 'Carol' specifically could not help but remind me of glamorous girl-friends since glamour as Carol practised it was seen as the means of managing both escape and a stigmatised identity (cf. Skeggs, 1997). Carol's public 'face' of 'ad girl', was the apparent opposite of my stance at her age when I took up the position of a classic and extraordinarily conformist 'good girl'. But this seemed to be the only possible position from which to make the move into further and higher education. So being back with Carol reminded me of exactly those splits I felt at the time between 'respectable' and the 'hard' working-class forms of femininity policed so ferociously by brothers and fathers – positions lived out as pro- or anti-schooling classed identifications.

The return into schools thus prompted intense moments of recognition – recollections of earlier painful times – of having to keep in with the same 'bad girls' at my secondary modern school. Given that I was aspiring and given that I equated aspiration with 'being good' and since being 'good' was seen as 'creeping' and 'being a swot' and being 'sexless', I realized just how much is repressed as painful in these class and gender lessons of academic, sexual and social modesty. These memories haunted negotiating and establishing under-standings with and about Carol. The differences between us were as uncanny as the similarities. As a prefect at school I had tactfully avoided the 'bad girls'' corner of the playground – here plotting and smoking prevailed and I learned not to put myself in the position of the class 'nark'. I managed to occupy that desperately compromised middle position – liked by both teachers and inoffensively 'popular' amongst my peers. Fieldwork had similar dynamics. I was careful not to court vulgar 'sameness' like some ill-fated 'trendy' teacher but tried to keep 'in touch' or 'with-it' enough in order to keep hold of the fast-swirling currents of girls' social popularity stakes.

What is important here is the simultaneous play of difference as well as similarities – to see more specifically that forms of being working class are split (see Reay, 1997). Indeed, I have shown how Carol's own sexual self was divided into contrasting public and private representations (Hey, 1997a, chapter 6). My own split positioning as 'sexless' was not only because femininity and the rational are counterposed (Walkerdine, 1984) but because the denial of any form of sexuality was lived as the only guarantee of surviving into higher education.

It is not possible to assess how much these (troubling) empathies and differences impacted upon the ways I have chosen to understand any of the girls in my study. The point is more that of acknowledging the density of points of connection and rupture that lie beyond the compulsions of identity, identification or boundaries of the self and the other. The claim is that we need to recognise the dynamic partiality of rapport and partial recognitions.

PARTIAL IDENTIFICATIONS: RELATIONS OF SELF/ OTHER : SKIVVIES, HUSSIES AND ALL STARS

Getting to know different girls included absorbing a sense of the girls' distinctive home and school lives. These differences were plotted along a continuum of onerous domestic duties (working-class girls) or demanding extra-curricular schedules (middle-class girls). The former demands signalled working-class girls taken-for-granted incorporation into the domestic division of labour, whilst the latter assumed an altogether more public agenda determined around accumulating, spending and replenishing cultural, symbolic and social capital (Reay, 1998). These classed patterns affected how I related to certain girls in the field. The fact that I could at one point feel more emotionally partial to the working class girls than at another more sympathy for the middle class girls, undoubtedly drew from my own history, biography and subsequent re-positioning (Hey, 1997b). How one begins to theorise the weight of these inter-subjective and intra-subjective meanings is the question I am interested in.

It is in these complex intricacies of same/difference dimensions that animate social interactions more generally but which under the aegis of self-conscious research relations become the material out of which we construe our understandings. In focusing down on one such set of relations I have selected out the network of meanings that I spun between Carol's present and my past. What I suggest is that these articulated (partial) recognition and (partial) understanding that provided the base for our rapport. Evidence from other writers (Cohen, 1997; Raphael Reed, 1997; Reay, 1998; Skeggs, 1997) indicates a welcome recognition that auto/biographical traces are worthy of critical analysis.

TROUBLING IDENTIFICATION/S: SKEGGS ON SAMENESS

Beverley Skeggs refers to Ortner's comment, that in anthropology there are extensive debates about how we construct 'others' but she asks 'who produces us?' Skeggs responds:

I brought histories, locations and identifications to the research. These informed the interpretations I made . . . Reluctant as I was to become interested in myself, questions of who are 'they' informed who 'I' was (Skeggs, 1997, p. 34).

She then goes on to note the shifting nature of both her affinity with and her difference from the working class women she studied. Increasingly, earlier identifications were cross cut by her increasing alienation as she was absorbed into :

. . . an alternative value system which protected me from the pressures they experienced . . . Every time I walked away from a visit in the later stages of the research, I experienced a sense of physical and metaphorical escape; *it could have been me* (Skeggs, 1997, emphasis added).

Skegg's move anticipates a model of identification that does not collapse into uncritical empathy nor cosy bogus humanitarian 'sameness'. What she shows instead is the outcome of researching as a 'shifter in position' (Cohen, 1997). Valerie Walkerdine theorizing the construction of the 'fictions' of identity has, like Skeggs and others (see Zmroczek & Mahony, 1997), noted the psychic costs of social and psychological transfers across class and gender categories (Walkerdine, 1990; cf. Hill-Collins, 1998). In fieldwork we can re-live these shifts or 'transpositions' (Cohen, 1997) as embodied. I frequently experienced prolonged research exchanges with white working class school girls accompanied by feelings of intense *deja vu* – "it could have been me" (Skeggs, 1998). These modes of recognition are too powerful to be conveniently dismissed as subjectivist delusions that 'get in the way' of the 'real' business. They need *more* not less analytical attention.

I am mindful that whilst research is not a proxy for self-analysis, these moments of 'return' speak to an intelligent apprehension of the 'other' reliant upon a reanimation of one or other 'former' selves. In this case to previous modes of cultural capital buried beneath middle class acculturation (cf. Raphael Reed, 1997). Phil Cohen conceptualises 'others' marks upon our psyche as a 'hidden curriculum vitae', formed by the projected desires of powerful others (Cohen, 1999). I suggest that another possibility is to see our hidden curriculum vitae as vital self-production – a resistant act towards those who would erase the difference of 'past lives' (Hey, 1997b). In the next section I pursue this reading in a more systematic and reflexive way.

IT COULD HAVE BEEN ME: PSYCHIC RELATIONS OF RESEARCH

My research into girls included mapping close as well as antagonistic relations. A common theme marking all of the girls relations was a concern with finding a comfortable place within the large and small differences that made up their social relations. Their relationships therefore involved intense identifications with and against other girls. Girls in my study made sense of themselves as inexorably bound up with the 'other' as both a relation of fear and loathing and desire:

> Girls *have* to make sense of themselves *against other girls* but they have to do so "not in conditions of their own choosing." We can I think locate some of the features of girls' relations here. We have seen that girls 'longings' for certain girls; for a sense of belonging to certain groups and argued that these affinities resonate as another politics of 'desire' played out in the in/ex/clusions of personal forms of feminine intimacies (cf. Steedman, 1986, p. 33). There is however, more to it than that. Not only were 'places' desired, they were loathed, not only wanted they were repudiated. Moreover, given that the 'places' were embodied by 'other' girls and all they represented – looks, clothes, manners, forms of sexual self display or 'cleverness' – we should not be too surprised to discover that the various economies of girls' friendships carried both intense sources of personal affiliation as well as forms of social antagonism (Hey, 1997a, p. 136).

It now occurs to me that these processes of struggle born of contradictions inhere to some extent in research relations. The call for 'rigour' in research is often over-determined by the need to appear robust in the face of desire and difference. However, I am not sliding over into gross subjectivism. My own position is a paradoxical, not to say unusual one, of psychic pragmatism! In theorising rapport it seems to me that we may have no choice but to accept that social relations also trace the uneven 'unknowable' power of the psychic, intuitive and autobiographic. If we accept this premise, then it is possible to see that sometimes rapport is driven by varying degrees of desires for: connection; communication; translation; mutuality and understanding. These feelings are played out in the yearning towards the 'other' who may be both at the limit conditions of current 'difference' but who may also contradictorily provoke re-discoveries of displaced affinities. Such loaded research exchanges can be lived as a series of imaginary returns to surprisingly familiar social milieus.

As feminist researchers we may have impeccable credentials about equalizing relations, of operating non-hierarchically yet we live enmeshed within those same relations of power we are commenting upon and seeking to change. Whilst this is acknowledged (see Haraway, 1988), the emotional concomitants of this positioning are unspoken. Yet for me, at times, fieldwork is best understood as both a socio-emotional as well as a rational economy – full of movements to and against an 'other' which are lived as a complex of empathy, indifference and hostility in other words a struggle (cf. Reay, 1998). Certainly in the case of my own work when it felt like so many of the girls 'could have been me', issues of the 'autobiography of the question' cast their long shadow. This was true of fieldwork and the interpretative process of re/ presenting girls' experiences. Saying this is not the same as grounding the project in mindless subjectivism. The move is more about critical subjectivism – reconceptualizing rapport sociologically as a relation of desire without conceding it as a guarantor of the truth or indeed as the main substantive focus of any study. Rather there is a need to recognize the psychic economy of rapport that should encourage an increased reflexivity about field relations which 'do not work', interviews which stall or research agendas that are simply undoable (see Hey, 1999b).

Few of us admit, let alone write about these unsuccessful encounters. When was the last time anybody admitted disliking anyone on fieldwork? Exceptions are the discussions by Diane Wolf (Wolf, 1996). This does not mean a retreat into 'elective affinities' as the answer (see Cohen's critique 1997); nor the privileging of researcher over the researched, but it might mean our asking deeper questions about what is happening when we do successfully work the space between 'us' and 'them'. For those of us who are 'shifters in positions', we can, I have suggested, push the psychic membrane further, to theorize the construction of 'rapport' as a possible example of a temporary stabilization of split identifications – a hegemonic moment.

If we do this we may find it useful to consider our investment in qualitative research in part attributable to the opportunities it provides for connection, communication, mutuality and understanding. More speculatively we can suggest that at times such desires may have a psychological impetus to square the proverbial identity circle. This feeling of completion may be imaginatively realised in the social and emotional capacity to recognise the 'other' as a proxy 'former' self. Then rapport could contain that (unspeakable?) desire for healing the psychic (social?) splits wrought in fractured lives. Richard Johnson explores similar difficult territory in his discussion of the impact of the death of Diana in terms of reactivating his prior experiences of mourning (Johnson, 1999). Certainly in the flux of field encounters and their interpretation I experienced an equivalent order of emotional resonance. My understanding of who I once was and who I had 'become' were cut through by powerful tugs of memory; of imagined or remembered instances when what girls were telling me spun me back into my own past – into classrooms and playgrounds redolent of similar pleasures and pains. This time round however I was differently armed. Whilst autobiography and the praxis of rapport enabled me both to be 'up close and personal' theory shifted me onto more distant and analytical ground. After all, it was through an education in theory that I had 'escaped' into the privileged world of intellectual labour – the world of sociology and women's studies. This was a world away from 'Carol's' mode of 'escape'. Yet, it was precisely my realization of having been similarly positioned by the discourse of schooling and respectability that led me not to want 'theory' to gloss over 'experience'. Neither, paradoxically did I want rapport as socio-cultural work to remain untheorized.

Exploring the layers of fieldwork relations allows us to contribute to discussions about the construction of identifications (Hall, 1996). We need to keep a focus on the contradictory nature of subjectivity with its psychic as well as its social dimensions. Why is it the case beyond the purchase of 'interests' that we invest in some discursive positions and not others. The theoretical and empirical elaboration of how identifications are secured (and broken) seems to me an urgent task for social investigators. We need to make this conceptual move to avoid concerns about discursive determinism (see and cf. Raphael Reed, 1997; Skeggs, 1997). Similarly, viewing rapport as dialectical addresses the question about partiality in research. Fine poses it thus: "few accounts [. . .] show that we do this work, much less *how* we do this work" (Fine, 1994, p. 22). A recognition of the 'other scenes' of research (after Cohen, 1997) that is an acceptance that social relations are spun out of history is one possible response to this challenge.

SOME CONCLUSIONS: THE POWER OF EXPERIENCE – THE EXPERIENCE OF POWER

Rapport in this new poststructuralist theorization, can be understood as working through the multiple levels of experience articulated by Caroline Ramazanoglu & Janet Holland (1997). Their discussion about the 'levels of experience' is extremely helpful. They propose theorizing experience along the vectors of rationality and irrationality – the material and the discursive. Their account of experience incorporates events, the social, embodiment, emotions, values, ethics, theory, language, discourse, meanings construed through communication. Such a broad-based definition allows us to see the plural sites for the play of power; sites that are public, private, personal and political. They see research as located in these multiple dimensions. This adds to feminist understanding that already seeks to ameliorate institutionalized power held between individuals, through a conscious effort at avoiding assertions and practices that seek to position the 'other' as inferior, subordinate or problematic.

Moreover, at an individual level within these wider political aspirations rapport can for those who have shifted position be experienced (problematically) through psychic commitments to an 'imagined' 'desired' healing for the split self. Seeing rapport as a *practice* renders visible the complexity of fieldwork that relies on a great deal of *conscious* as well as *unconscious* adjudication. Constructing empathy involves us in sensitive and simultaneous negotiations across the social, cultural, psychic, emotional and rational. Rapport certainly demands simultaneously self-censorship and self-understanding. It is active even as it works in suppressing knowledge and social values that you did not know you even had. It is the result of actions that are both conscious and unconscious, the former framed by social, intellectual, personal moral choices conveyed in dialogues that are intellectually and emotionally exhausting.

This is why fieldwork, when you actually appear to be 'doing nothing' with responsibilities only to research and observe, is one of the most emotionally intense forms of social-intellectual interactional work. One false step can put your social research relations under jeopardy since social trust can dissolve in the trace of a voice shift; by your being in the wrong place or by often involuntary moves of the body indicating disapproval. These social glitches are as Hoggart and others remind us seldom innocent (especially in English) because they are indexical of all those micro-social relations of power we live within (Hoggart, 1990). Moreover, some of these communication cues are unconscious.

Within the micro-negotiation of social (research) relations appear the sorts of possible 'selves' disclosed in the discourses. There can be no unmediated narratives in this space. We cannot assume that, if 'like' interviews 'like', for example if sexual, 'racial', class or gender positions are 'matched' through 'elective affinities' that difference is controlled for or power equalized or that we will be in a better position to 'get at the truth'. There is always another difference beyond our reach to test out trust, mutuality, communication and meaning-making. Moments of field worker 'rapport' are wrested from and structured by such tensions which even the most seductive fieldworker is incapable of transcending. The idea of rapport as foreclosure is too simplistic. Rapport is better considered as radically unstable. It is a provisional condition that has to be won against the grain of multiple forms of difference. In that sense it is better seen as a temporary interpersonal alliance that can at times draw on imaginative identifications of autobiography as well as of current differences. My metaphor for rapport compares it to the sought-after hegemony of political alliances – the temporary contract between people that holds both 'sameness' and 'difference' in play and thus allows for action and representation.

As the above metaphor indicates I have a wider agenda to stake out here. One that, as I have indicated, draws from my own experience of doing fieldwork but one that locates the reconsideration of rapport within broader discussions about the need to explore the educational politics of subjectivity (Cohen, 1997; Walkerdine, 1987) and the social conditions of constructing dialogues in political communities. My own work on girls' alliances has shown how im/possible girls' cultures of friendship are – how worked through by relays of power and powerlessness performed through the dialectic of same/different/self/other. In sum, girls' solidarity was undercut despite their shared oppression in the face of masculine hegemonic discourse. Sexual reputation was still paramount, stage-managed and reworked by girls desperate to define themselves through real and imagined differences as the right side of difference – i.e. 'good girls'. Bernstein notes how power is always present in the policing of boundaries between categories (Bernstein, 1996).

Poststructuralist and feminist accounts persistently point to the need for directing more analytic attention to theorising what we have invested in sustaining difference. Celebration or retreat on the grounds of postmodern scruples can all too easily slide into self-regarding political inertia. I prefer a more engaged and thus more difficult intellectual politics. Part of this implies a form of research grounded in a politics of recognition of the 'other' as ineluctably bound up with the 'same' (see Spivak, 1990). My own efforts at deconstructing rapport derive from this political concern. They are also impelled by emerging work on class and subjectivity (Reay, 1998; Skeggs, 1997; Zmroczek & Mahony, 1997).

Despite real dangers and difficulties in "exploring the unconscious as a pedagogic and research practice," I see such a move as potentially richer and certainly more productive than the logics of postmodernism. In part I choose to engage this debate because the hyper-rational is now hegemonic in education at almost all levels (Kenway et al., 1997; Kenway, 1998; Slee et al., 1998). A commitment to exploring the relations and the articulations of same and difference stands against dominant knowledge modes as a materialist practice located in experience. I agree with Ramazanoglu and Holland that feminism must be located in 'experience'.

Experience we can communicate is like no other knowledge of what power relations exist
and how they are organised. This is not the only knowledge we need but experience ties
feminists to making sense of the levels at which people, in their diversity, are living out
their commonalties, contradictions and divisions. This stops us from escaping from the real
divisions between women (for example by relativising them as socially constituted 'truths'
or abstracted 'difference' and makes us consider the diversity of women's relationships.
(Ramazanoglu & Holland, 1997, p. 14)

In claiming back the risky project of 'communicating' the same and the 'other,
difference, diversity and commonalities', they heed the tyrannies of endless
deconstruction, reminding us what came into being as a movement against
injustice and oppression. As their history recounts, feminism was agentive in
challenging power relations, by deeds and words, through specifying the
material conditions of subordination and domination. Now such activism is
positioned as naïve but the possibilities of action can be rethought. There is, it
seems to me, an analogy of another 'in-between' space made in the process of
constructing or seeking to construct rapport. It is a political practice that might
need to borrow the 'humanist subject' as part of what Spivak calls 'strategic
essentialism' in order to give us the possibility of elaborating 'common'
humanity. Such recognitions are indispensable in mandating moral concerns
through identification in the other of the self (Bauman, 1990). It is after all the
splitting/failure of recognition of the other and the intense obscene and too
ready fear and pathologization of difference that produces the splittings of
fascism, misogyny, racism, and homophobia.

Building upon the critical insights of post-structuralism and post-modern-
ism, though not imprisoned by them, feminist scholars in association with
others have a responsibility to continue to 'dream of a common language' or
'practical utopias'. Taking back or taking up the common sense rhetoric of
emotional and intellectual critical reflexivity (Giddens, 1993) seems to me one
opportunity. There is plenty of evidence that the desire for belonging – people's
thirst for emotional connection – is endless and is not invariably reactionary as
many on the left and many post-modernists believe. Along with others, I argue
that desires (for be/longing) are not inherently anything it depends on the
contexts within which these urges emerge – the meanings and longings they
carry and are made to carry – the purposes to which they are put and what they
mobilize around (see Hey, 1999c).

In my own work I have sought to theorize the ways in which codes of
oppression and desire co-construct each other – so that 'race and reputation';
sexuality and 'respectability' co-mingle within the discursive framings of girls'
cultural and social subordination. Girls "in many ways do the work of
subordination themselves" in putting each other into desired and undesired

places. (Hey, 1997a). Such a position reflects a refusal to fall into the impossible indeterminacy of difference as the new iron law of the 'other' since these differences are material and hierarchical. They carry different forms of social, economic, cultural, symbolic and emotional capital (Reay, 1998). Yet this position still holds simultaneously onto an understanding that sees the relentless inter-dependency of these differences as modes of inter-subjectivity.

Stuart Hall's move to reconceptualize a non-essentialist notion of (racial) desiring/difference as 'the product of' 'distributive relations' captures the dialectical 'economy of othering and saming' that I am striving to describe:

> Contrary to the superficial evidence, there is nothing simple about the structures and dynamics of racism . . . It is racism's very rigidity that is due to its complexity. Its capacity to punctuate the universe into two great opposites masks something else; it masks the complexes of feelings and attitudes, beliefs and conceptions, that are always refusing to be so neatly stabilized and fixed . . . All the symbolic and narrative energy and work is directed to secure us 'over here' and them 'over there', to fix each in its appointed species place. It is a way of masking how deeply our histories actually intertwine and interpenetrate; how necessary 'the Other' is to our own sense of identity; how even the dominant, colonizing, imperialising power only knows who and what it is and can only experience the pleasure of its own power of domination in and through the construction of the Other (Hall, 1992, p. 16).[2]

If we come to recognize ourselves through difference and desire – rapport (like love?) would seem an alternative quest to work against the logics of separation to re/dis/cover selves in the 'same'. My initial interest in theorizing 'rapport' spills over from my own immediate interest in thinking about field relations, to bolder claims that we now need to pay far more analytical as well as political attention to the 'same'. This interest connects up to the work of others such as Michelle Fine, through her call for 'transcoding', Homi Bhabba's text based insistence on 'translations' as a 'third space' (Bhabba, 1990) as well as to concerns with renewing civic society in bids Nira Yuval-Davies makes to seek possibilities for 'transversal dialogue' (Yuval-Davies, 1990). The emergence of these cognate strategies suggests compound moves to reinstate the possibility of collective or at least collaborative or cooperative 'action' in contrast to a withdrawal from taking *any* responsibility for anybody on the post/modern grounds that nobody can claim to speak for anybody about anything.

BEYOND BINARIES: THIRD SPACES AND OTHER PLACES

In the feminist materialist position the motion and indeed the possibility of community is core to the workability of an accountable feminist research. It is a model that is premised on a reflexive knowing or epistemic community –

"collectivities needing to interrogate our constitution as collectivities." Drawing on Code (1991), Ramazanoglu and Holland (1997) do not lie down in front of difference but instead argue that:

> The construction of knowledge is an intersubjective process, dependent for its achievement on communal standards of legitimation in the power and institutional structures of communities and social orders (Code, 1991, p. 132 cited in Ramazanoglu and Holland 1997, p. 13).

To the extent that others recognize themselves (or not)[3] in my text, then I too would, as I have argued, see myself as engaged in such a project of seeking validity from an epistemic community. Spivak calls for a way of moving beyond binaries to "trace the other in the self" (Spivak, 1990, p. 326); this call echoes the argument of this chapter. At another level such moves to see sameness offer the potentiality for engaging difference from a more optimistic basis.

To the extent that feminist language has excluded the experiences of 'others' it has found itself open to non-validation by challenge from different communities of black, working class, older and disabled women. As a collective project, it has had to continuously re-invent itself through the terms of these exclusions and re-think the force of how new knowledge re-positions prior knowledge. To the extent that recent currents in academic feminism have moved to see themselves as more accountable to the intellectual border patrols of postmodernism, than 'others' in their 'differences' it is reactionary. The new postmodern politics of accountability appear to have shifted the politics of knowledge production away from experience (in all its multiple layers and complexities) back to those who assert (pace Foucault, 1972) that knowing is 'violence' and 'colonization'. But knowing has the potential to empower as well as colonize. It depends on more than the discursive – it implicates knowers' motives, the social orders of power within which the knowing takes place and the politics of accountability.

Power cannot be appeased primarily if at all by practices of abstract intellectualisation. What is required is a different sort of risk-taking – one of taking up (invariably normative) positions through seeking empirical investigations of real people in recognizable situations. The rush to distance from the 'other' is I think closing down of the responsibility for knowing and knowledge production of the self and 'other'. The really difficult challenge is to work the space between us/them/you/me/self/other and to articulate points of connection and disconnection as these are secured in relations of power, in order to argue for their change. Not to do so merely secures the prevailing hierarchies.

I have been interested in this chapter in showing that research endeavours are invariably compromised by the material conditions of power. In this respect, data collection and interpretation transpire in the overlapping contexts of: the current state of the field; the degree of self-understanding of the researcher including those cultural capacities to read beneath the said to theorize the silences and antagonism(s). My own and others' work is conceived fundamentally against the new methodological purism of post-positivism (cf. Ramazanoglu, 1992). This is not to imply that it is complicit with a too glib notion of feminist activism (Hey, 1997a). It is for a critical materialist feminism that wants to change the world, or more immediately the educational and social experience of girls. This latter position occupies the messy third in-between space of a materialist deconstructivism.

NOTES

1. It occurs to me that I would have been far less able to show the mundane interpersonal talents demanded by fieldwork had I not been able to call upon the cultural capital of growing up working class. This is not to claim that only working-class researchers can study the working class! Or that there is one way to be working class. But it is to acknowledge the effects of personal experience on social research. I would I think have been less sensitized to what was at stake in girls' classed cultural re/solutions had I not had broken some of their regulations.

2. There is a specific history of racism(s) which cannot be appropriated by analogy to other forms of power relations but Hall's model does offer a purchase on the psychic and social interdependencies of many self/other formations.

3. I fully acknowledge that my book does not problematize either my own whiteness sufficiently, nor does it engage in the depth of analysis of the ways in which the ethnicity of the girls I studied was specifically implicated in their social positioning around sexuality and social class. This point was amplified by a very helpful and generative review of my book by Mairtin Mac an Ghaill (1998).

REFERENCES

Bauman, Z. (1990). Effacing the Face: On the Social Management of Moral Proximity. *Theory, Culture and Society, 7*, 5–38.

Bernstein, B. (1996). *Pedagogy Symbolic Control and Identity: Theory, Research, Critique.* London: Taylor and Francis.

Bhabba, H, (1990). Interview with Home Bhabba: 'The Third Space'. In: J. Rutherford (Ed.), *Identity, Community, Culture, Difference.* London: Lawrence and Wishart.

Britzman, D. (1995). The Question of Belief: Writing Post-structural Ethnography. *Qualitative Studies in Education, 83*, 229–238.

Cohen, P. (1997). Re-thinking the Youth Question. In: *Re-thinking the Youth Question – Education, Labour and Cultural Studies.* London: Macmillan.

Cohen, P. (1999). Autobiography and the Hidden Curriculum Vitae. In: P. Cohen (Ed.), *Studies in Learning Regeneration.* Centre for New Ethnicities Research: University of East London.

Code, L. (1991). *What Can She Know? Feminist Theory and the Construction of Knowledge.* Ithaca and London: Cornell University Press.

Crawford, J., Kippax, S., Onyx, J., Gault, U., & Benton, P. (1992). *Emotion and Gender: Constructing Meaning from Memory.* London: Sage.

Davies, B. (1993). Beyond Dualism and Towards Multiple Subjectivities. In: L. K. Christian-Smith (Ed.), *Texts of Desire: Essays on Fiction, Femininity and Schooling.* London: Falmer Press.

Finch, J. (1984). 'It's Great to Have Someone to Talk to'. The Ethics and Politics of Interviewing Women. In: C. Bell & H. Roberts (Eds), *Politics, Problems and Practice.* London: Routledge.

Fine, M. (1994). Dis-stance and Other Stances: Negotiations of Power Inside Feminist Research. In: A. Gitlin (Ed.), *Power and Method: Political Activism and Educational Research.* London: Routledge.

Foucault, M. (1972). *The Order of Things: An Archaeology of the Human Sciences*, trans. A. M. Sheridan-Smith. New York: Pantheon.

Giddens, A. (1993). *The Transformation of Intimacy: Sexuality, Love and Eroticism in Western Societies.* Cambridge: Polity Press.

Hall, S. (1992). Race, Culture and Communications: Looking Backward and Forward at Cultural Studies. *Rethinking Marxism, 5,* 10–18.

Hall, S. (1996). Who Needs Identity? In: S. Hall and P. du Gay (Ed.), *Questions of Cultural Identity.* London: Sage.

Haraway, D. (1988). The Science Question in Feminism and the Privilege of Partial Perspective. *Feminist Studies, 14*(3), 575–597.

Hey, V. (1982). The Necessity of Romance. *University of Kent Occasional Paper 3.* Canterbury, University of Kent.

Hey, V. (1997a). *The Company she Keeps: An Ethnography of Girls' Friendships.* Buckingham: Open University Press.

Hey, V. (1997b). Northern Accents: Southern Comforts: Subjectivity and Social Class. In: C. Zmroczek & P. Mahony (Eds), *Class Matters: 'Working Class Women' and Social Class.* London: Taylor and Francis.

Hey, V. (1999a). Reading the Community: a Critique of some Postmodern Narratives of Citizenship and Community. In: P. Bagguley & J. Hearn (Eds), *Transforming Politics: Power and Resistance.* Basingstoke: Macmillan.

Hey, V. (1999b). Frail Elderly People: Difficult Questions and Awkward Answers. In: S. Hood, D. Mayall & S. Oliver (Eds), *Critical Issues in Social Research: Power and Prejudice.* Buckingham: Open University Press.

Hey, V. (1999c). Be(long)ing: New Labour, New Britain: the 'Dianization' of Politics. In: A. Kear & D. L. Steinberg (Eds), *Mourning Diana: Nation, Culture and the Performance of Grief.* London: Routledge.

Hill-Collins, P. (1998). Address to AERA April (1998).

Hoggart, R. (1990). *Life and Times Volume 1: A Local Habitation.* Oxford: Oxford University Press.

Johnson, R. (1999). Exemplary Differences: Mourning (and Not Mourning) a Princess. In: A. Kear & D. L. Steinberg (Eds), *Mourning Diana: Nation, Culture and the Performance of Grief.* London: Routledge.

Kehily, M. (1995). Self-narration, Autobiography and Identity Construction. *Gender in Education, 7*(1), 22–31.

Kenway, J., & Willis S. with Blackmore, J., & Rennie, L. (1997). *Answering Back: Girls, Boys and Feminism in Schools*. St Leonards, New South Wales: Allen and Unwin.

Kenway, J. (1998). *Local/Global Labour Markets and the Restructuring of Gender, Schooling and Work*. Paper presented at the American Association for Research in Education conference, San Diego, April International symposium on Gender, Education and Globalisation.

Kenway, J. with Langmead, D. forthcoming. Fast Capitalism, Fast Feminism and Some Fast Food for Thought. In: S. Ali, K. Coate & W. wa Goro (Eds), *Belonging: Contemporary Feminist Writing on Global Change*. London: University College London Press.

Kuhn, A. (1995). *Family Secrets: Acts of Memory and Imagination*. London: Verso.

Lather, P. (1998). *Against Empathy, Voice and Authenticity*. Paper presented at American Educational Research Association, San Diego April.

Lather, P., & Smithies, C. (1998). *Troubling the Angels: Women Living with HIV/AIDS*. Boulder, Westview: HarperCollins.

Mac an Ghaill, M. (1998). Review of *The Company She Keeps*. In Review Symposium *British Journal of Sociology of Education, 19*(1), 135–142.

Mercer, K. (1990). Welcome to the Jungle: Identity and Diversity in Postmodern Politics. In: *Identity, Community, Culture, Difference*. London: Lawrence and Wishart.

Morrison, T. (1992). *Playing in the Dark: Whiteness and the Literary Imagination*. Cambridge MA.: Harvard University Press.

Oakley, A. (1981). Interviewing Women: a Contradiction in Terms. In: H. Roberts (Ed.), *Doing Feminist Research*. London: Routledge and Kegan Paul.

Ramazanoglu, C. (1992). On Feminist Methodology: Male Reason Versus Female Empowerment. *Sociology, 26*(2), 207–212.

Ramazanoglu, C., & Holland, J. (1997). *Tripping over Experience: Some Problems in Feminist Epistemology*. Paper presented at the Transformations: Thinking through Feminism Conference, Lancaster University 17–19 July.

Raphael Reed, L. (1997). *Researching: Re-finding and Re-making. Exploring the Unconscious as a Pedagogy and Research Practice*. Paper presented at the BERA conference, Educational Research in Britain: Power and Method.

Reay, D. (1997). The Double-bind of the 'Working-class Feminist Academic. The Failure of Success or the Success of Failure. In: C. Zmroczek & P. Mahony (Eds), *Class Matters: 'Working Class Women' and Social Class*. London: Taylor and Francis.

Reay, D. (1998). *Class Work: Mothers' Involvement in their Children's Schooling*. London: University College Press.

Sherratt, N. (1983). Girls, Jobs and Glamour. *Feminist Review, 15*, 47–60.

Sheurich, J. J. (1997). *Research Method in the Postmodern*. London: Falmer Press.

Skeggs, B. (1998). *Formations of Class and Gender: Becoming Respectable*. London: Sage.

Slee, R., Weiner, G. with Tomlinson, S. (Eds) (1998). *School Effectiveness for Whom? Challenges to the School Effectiveness and School Improvement Movements*. London: Falmer Press.

Spivak, G. (1990). *The Post-Colonial Critic: Interviews, Strategies, Dialogues*. London: Routledge.

Stacey, J. (1988). Can There be a Feminist Ethnography? *Womens' Studies International Forum, 11*(1), 1–28.

Steedman, C. (1982). *The Tidy House: Little Girls Writing*. London: Virago.

Steedman, C. (1986). *Landscapes of a Good Woman: A Story of Two Lives*. London: Virago.

Walkerdine, V. (1984). Developmental Psychology and the Child-centred Pedagogy: the Insertion of Piaget into Early Education. In: J. Henriques, W. Hollway, C. Urwin & V. Walkerdine

(Eds), *Changing the Subject: Psychology, Social Regulation and Subjectivity*. London: Methuen.

Walkerdine, V. (1985). Dreams from an Ordinary Childhood. In: L. Heron (Ed.). *Truth, Dare or Promise: Girls Growing Up in the Fifties*. London: Virago.

Walkerdine, V. (1985). On the Regulation of Speaking and Silence. In: C. Steedman, C. Urwin & V. Walkerdine (Eds), *Language, Gender and Childhood*. London: Routledge and Kegan Paul.

Walkerdine, V. (1987). Femininity as Performance. *Oxford Review of Education, 15*(3), 267–79.

Walkerdine, V. (1990). *Schoolgirl Fictions*. Verso: London.

Whitehead, A. (1976). Sexual Antagonism in Herefordshire. In: D. Barker & S. Allen (Eds), *Dependence and Exploitation in Work and Marriage*. London: Longman.

Wolf, D. L. (1996). Situating Feminist Dilemmas in Fieldwork. In: D. L. Wolf (Ed.), *Feminist Dilemmas in Fieldwork*. Boulder Colorado: Westview Press.

Yuval-Davies, N. (1990). Women, Citizenship and Difference. Background paper for the Conference on Women and Citizenship, University of Greenwich, London, July.

Zmroczek, C., & Mahony, P. (1997). *Class Matters: 'Working Class' Women and Social Class*. London: Taylor and Francis.

OPENING THE CAN OF WORMS: GENDER AND EMOTION IN SENSITIVE RESEARCH

Denise Carlyle

[The] ethnographic method exposes subjects to far greater danger and exploitation than do more positivist, abstract, and 'masculinist' research methods . . . the greater the intimacy . . . the greater the danger (Stacey, 1991, p. 114).

Researching sensitive issues with vulnerable populations invites "the disclosure of highly personal and confidential information" (Brannen, 1988, p. 552). Sensitive research "potentially poses a substantial threat to those who are or have been involved in it" and may present problems "because research into them involves potential costs to those involved in the research, including, on occasion, the researcher" (Lee, 1993, p. 4). Highly charged with emotion, the interview may be an intensely stressful experience for both interviewee and interviewer. Emotional pain is shared. Emotional taboos may be shattered.

To reach deeply into another's inner world holds great responsibilities, not only regarding accountability and protection toward the participant self but also the fieldworker self. In order to protect the self, emotion may be held defensively beneath the surface in the subconscious. That subconscious is the 'can', holding our 'shadow selves' (Abrams & Zweig, 1991). Exploring sensitive issues may incur opening that can, causing the resurfacing of uncomfortable, distressing feelings – the 'worms'.

Opening the 'can of worms' is potentially dangerous. While the interview may provide "a platform for people to speak their minds in a way, and in such detail that rarely occurs to the ordinary person," it may also prove to be "a

Genders and Sexualities in Educational Ethnography, Volume 3, pages 185–207.
Copyright © 2000 by Elsevier Science Inc.
All rights of reproduction in any form reserved.
ISBN: 0-7623-0738-2

cataclysmic 'critical' event for the interviewee, bringing about a redefinition of personal identity and aims" (Woods, 1986, pp. 69–70). It may prove so for the interviewer (Ely et al., 1991). Those in the dialogue may well be unprepared for the emotional and cognitive distress which may have to be dealt with both within and outside the interview experience. Reliving past emotion may precipitate emotional crisis with a subsequent need for immediate therapeutic intervention (Cowles, 1988).

Until recently, emotion has been relatively neglected both as a research topic, and as producer of, and product of, interview transactions. However, emotion plays a "central role in the human experience and cultural scripts of health, sickness, disability and death" (Williams & Bendelow, 1996, p. 47). To explore beyond common-sense accounts of emotional experience, to bring forth 'real' emotion versus 'script' emotion, requires that fieldworkers possess consider-able emotional literacy (Goleman, 1995). To keep both interviewer and interviewee safe requires a heightened awareness and understanding of emotion within the interview transaction and within personal worlds. Fieldworkers require a particular mix of 'skills' – the ability to listen to, identify, interpret and analyze differing emotional voices; to manage feelings and uncover dissonance; to share power and intimacy, to demonstrate vulnerability, to self-disclose. What voices do fieldworkers need to explore sensitive issues? What voices do they need to protect both those they interview, and themselves?

In today's complex world, we have potential access to a multiplicity of voices (Gagnon, 1992). Much has been written of the influence of the gendered voice (Warren, 1988). Warren concluded that "the focal gender myth of field research is the greater communicative skills and less threatening nature of the female fieldworker" (Warren, 1988, p. 64). However, other readings suggest that there may be gendered differences in emotional, empathic, and communicative skills. For James (1989) the gendered division of labour in both private and public spheres encourages women and men to develop differing emotional skills. There are claims that, in much of the Western world, restrictive emotionality results in many males: (a) finding it difficult to understand, deal with and express emotion; (b) revealing thoughts more than feelings; (c) making fewer self-revelations to other males and (d) disclosing more to females (Kilmartin, 1994). It would appear that women and men may communicate in different ways (Tannen, 1991).

What meanings does this hold for the sex/gender of the fieldworker? Was I able to succeed in this type of research solely because I am a woman? Reflections on the research process showed that while the gendered voice was important, I had access to many other voices. It was not the gendered voice *per se* which held most influence, but access to a wide voice repertoire, a voice repertoire which is accessible not only to some women but also to some men. Through the 'Worm Soufflé' and the voices of my interviewees – through Luke, Stephen, Andrew, Charlotte, Maureen, Rebecca, Harry, Marcus and Ralph – I explore gender and emotion in field relations.

THE WORM SOUFFLÉ

Never open the can of worms unless you are prepared to make a worm soufflé (Geraldine Bown, 1998).

I was prepared to open that can, to make that soufflé, where raw, indigestible 'worms' might be explored, and, if necessary, dealt with.

Fieldwork

For a PhD study I am researching the sensitive topic: "Emotion and stress-related illness and burnout among secondary school teachers." Stress-related illness results when constellations of chronic negative stressors, unrelieved by positive stressors, accumulate from the interaction of individual factors, work pressures, family pressures and environmental demands (Pearlin, 1989; Woods, 1995). Burnout (Schaufeli et al., 1993) has been described as "a terrible ordeal" and "a tragedy, accompanied by intense personal pain" (Graham, 1995, p. 1). My aim was to explore "human lived experience" (Ellis & Flaherty, 1992, p. 1). Through individual testimony of illness trajectories (Strauss, 1987) I aimed to provide detailed qualitative and sociological input into the area of stress, burnout, and teacher emotion. According to Woods (1985, p. 17) life histories "tune into the process and flux of life, with all its uncertainties, vicissitudes, inconsistencies and ambiguities, but on a deeper scale, for they reach the subjective realities, pull in the historic and contextualize the present within the total framework of individual lives."

I interviewed ten women and eleven men, between one and nine times over 19 months, averaging four sessions each. Eight were teaching, nine in alternative employment, four on extended sick leave. Twelve spouses (8 female, 4 male) agreed to tell their stories. Three teenaged children volunteered their experiences of the parent's stress career. Interviews took place mainly in the home, generally lasting two to three hours, with around one and a half hours being taped and later fully transcribed, analytical summaries being returned to interviewees for validation and additional comment.

I talked with those in the first stages of crisis, those in the throes of breakdown, and those striving to rebuild their lives. I perused photograph albums, delved into secluded corners, listened to intimacies and secrets and, on occasion, swapped confidences. In some cases I observed despair as personal and professional worlds disintegrated. With some others I witnessed the delights of self-regeneration, of identity reconstruction. I was asking these teachers to delve beneath the public face at a very painful stage in life, when many felt at their most vulnerable. There were many tearful occasions. Both men and women cried openly. I was deeply moved by many narratives, and was ready to make that visible to participants. I did not hide my feelings. Occasionally I too was close to tears. How did I deal with this?

I felt it morally necessary to have access to a safety net. I adopted a humanistic interviewing framework derived from person-centred counselling (Rogers, 1951; Mearns & Thorne 1988; Paterson 1997). I understood this as encouraging a facilitative relationship, most likely to provide protection for both conversational partners, the emphasis on mutual benefit offsetting fears of exploitation (Cannon, 1989). I aimed to create a safe 'good-enough' holding environment (Stapley, 1996), characterized by genuineness, empathic under-standing, and unconditional positive regard, where participants might both explore their personal worlds and contribute to some research they felt to be of potential value to their profession, where the interviewing role was both a "data-collecting instrument for the project and a data-collecting instrument for those whose lives were being researched" (Oakley, 1981, p. 49). We cannot ask unless we are prepared to give (Lather, 1986).

What was it I felt I might offer that perhaps others could not? I felt I had life experiences which would provide access to voices which would enable me to deal with the emotions which might surface in sensitive interviews. The experiences of fieldwork showed how exploring emotional worlds requires a complex knowledge and understanding of one's own and others' emotions plus skills in monitoring, appraisal and management (James & Gabe, 1996).

Soufflé Preparation: Acquiring the Qualities of the Soufflé Chef

While we all have the potential to develop many voices, experiences may nurture or impair their expression and realization (Francis, 1997). Reflections on my own life history give a brief insight into some of the voices befitting the task of soufflé chef. Voices developed through a wide range of personal experiences within:

- the familial sphere – mothering two boys, rebellious farming wife, caring for sandwich agricultural students, miscarriage, bereavement, near death experiences, and marital breakdown;

- the public sphere – various organizational cultures within teaching, the Citizen's Advice Bureau as volunteer and trainer, sixth-form psychology and GNVQ teaching, research associate introducing and facilitating quality improvement with primary health care teams (Hearnshaw et al., 1998), training in multi-disciplinary counselling, piano teaching, choral conducting and singing, amateur dramatics.

Being the soufflé chef was not easy. I do not view the process as a maturing of innate traits but rather as a struggle of negotiating gendered identities, which I explore through the voices of early life, marriage and academic life.

The Voices of Early Life

One is not born a woman: one becomes a woman (de Beauvoir, 1949, p. 295)

One also becomes a man (Mac an Ghaill, 1996). I was a tomboy, raised in a Scottish mining community. Like many of my peers, childhood and adolescence were a time of identity struggle as I came to terms with societal expectations of what it meant to be female and male (Sharpe, 1976; Stanley 1993). I internalized the feeling 'rules' of my community – "the socially shared guidelines, governing the extent, type and intensity of feeling" (Bulan et al., 1997, p. 237) – and the sanctions applied against non-conformity such as invisibility, intimidation, sexual harassment (Stanworth, 1983). I learned Kipling's Six Serving Men – the What, Why, When, How, Where and Who of holding and expressing emotion such as fear, anger, joy. According to Rogers et al., during adolescence, girls experience a loss in ego development as they actively struggle with the "debilitating conventions of female behaviour" (1994, p. 30). Through assessing the emotional climate within the family and the classroom, I avoided negative feelings through emotion management of the self. Masculine gender orientations were discouraged, feminine gender orientations encouraged. Certain voices such as the maternal, the empathic, came to the forefront (Karniol et al., 1998). Some other available voices, the intellectual, the instrumental, became silent, frozen in time. By the age of 18, like many of my peers, I had capitulated, giving up all ambitions of a career in the philanthropic sciences, opting for mate hunting and the safety of music teaching.

I also struggled with the voice of disability. A polio survivor, currently dealing with the Late Effects of Polio, I know what marginalization means for me, what it is to be the outsider. In childhood I experienced ostracization by unenlightened peers in primary school. As a disabled working-class female teenager in grammar school, I felt alien. As a mature woman, I continue to experience 'otherness' (Gornick, 1971). Stress-related illness often incurs marginality. The disabled voice and the feminine voice have given preparation for the task of exploring the emotions of marginality .

Feminist Voices

Marriage entailed becoming a full-time farming wife as my husband began a career in farm management on a large traditional estate. Rural life was alien to me. It was through the experiences of the rural economy that I developed a greater awareness and understanding of:

- the emotion management dichotomy – emotion prescribed and regulated by others on the one hand, and autonomous, regulated by the self, on the other (Tolich, 1993);
- emotive dissonance – conflicts between what we feel, what we think we should feel, what we want to feel and what we try to feel (Hochschild, 1983).

Farming dominated family life. I found the working and social milieu controlling and disempowering, any needs I might have completely subsumed beneath the needs of the estate. I began to lose any sense of autonomy. I could not 'tug the forelock'. I could not accept management giving me orders via my husband, providing me no access to negotiation, or indeed refusal. Friends and family ('townies'!) enthused about the joys of country life, assuming I must find it idyllic, while I felt isolated and unhappy, without a supportive social network. I tried hard to play the farming wife but did not succeed. Resentment, frustration, powerlessness and anger simmered within.

I turned to raising two male children – a fascinating journey for any embryonic feminist. I watched their choice of voices narrow in response to their social world. It was near impossible to employ non-sexist childrearing methods (Statham, 1986). I was often told I was inviting future problems by making them different. Hochschild (1997, p. 210) proposes that one of the main skills which develops within the family is "the ability to forge, deepen and repair family relationships" which entails "noticing, acknowledging and empathizing with the feelings of family members, patching up quarrels, and soothing relationships." It is more likely that mothers have the opportunities and time to develop such skills. According to Eichenbaum & Orbach (1983), women tend to become responsible for emotional labouring as they constantly make compromises between their own needs and those of others. I learned to cope with children's fears, their distress and their tears, trying to ease their way in the world. I provided nurturance, comfort and protection. This I experienced as empowering. However, harnessing such skills can also be strategic. Through monitoring, assessing and shifting the focus of family encounters, I achieved instrumental ends such as creating harmony or conflict (O'Brien, 1994). Through the use of rewards and sanctions, I learned to use emotion as a means of ordering people, as a mechanism of social control (Miller, 1991). The perceived inequities of farming life led to the development of feminist voices. This was a voice of injustice and anger. 'Taming' these feminist voices led to the return of intellectual and instrumental voices.

Academic Voices

Through the WEA and the Open University I acquired an academic background in women's studies, psychology and sociology. Politically and academically, feminism became central to my life. I learned to hide this voice from unsympathetic peers. I studied counselling, family therapy, marital therapy and transactional analysis. The emotions which resurfaced here, the aftermath of childhood polio, were difficult to handle, necessitating my receiving counselling. Gaining funding for a full-time Masters degree in Interdisciplinary Women's Studies at Warwick, my focus became women's mental health, interviewing 'survivors' of invasive psychiatry, in particular electro-convulsive therapy. With little awareness, understanding and experience of feminist interviewing, the 'can of worms' or the 'worm soufflé', I was allowed naively to enter the lives of six women for a single interview. Meeting one woman many months later, I discovered I had unwittingly opened her can, my questions resulting in a traumatic reassessment of work and family life. Study for the PhD has given me a greater grasp of the perils of qualitative interviewing and the protections required for all parties. It is only now that I feel I have amassed the necessary skills to accomplish this type of interviewing. The feminist voice, in combination with instrumental, expressive, counselling and gendered voices, now has an emotional and cognitive balance appropriate to the task of soufflé chef.

Soufflé Baking: Enabling disclosure

The Influence of the Gendered Voice

In what ways did being female influence disclosure? Arendell (1997, p. 348) writes, "I came to understand that it was not me so much as a person having a particular interactional and interview style to whom they were sharing their stories. Rather, they were relating to me on the basis of their expectations of me as a woman." Being positioned as a woman (Davies & Harré, 1990) and having access to the feminine voice gave advantages in conducting this research.

For some women it was important to be interviewed by another woman. Rebecca was the victim of a sexual assault by a pupil, which brought a resurgence of emotional trauma from childhood. She described conversations with her female psychiatrist: "it was easy to talk to her because she was a woman . . . it's easier to talk to women about it even though the men have been wonderful. They really understand more what it means than men, however supportive they feel." From observations of marriage, family life and working environments, Rebecca believes that "women know their emotions . . . women tend to chat more and men tend to bottle it up more." Her description of a new partner gives insight into her perceptions of gendered conversational styles – "the first thing that struck me about him was that he listens very carefully to what you say, he talks effortlessly, which is so rare, not in all men, but it's so hard to find, isn't it, somebody who talks in depth and listens and thinks and takes it in."

Ralph explained his need for a female confidante, "I still don't think (sighs) you know that there is enough of a situation whereby male staff are able to talk to males. It's still perceived as a sign of weakness . . . I tend to go and talk to one of my female colleagues . . . partly because the two colleagues I would talk to are still so bound up on the treadmill that they don't really have time for that sort of conversation. They're too busy running still from meeting to meeting. Unable to stay and chat." He talked of a male colleague with marital troubles, "that sort of thing's so personal you won't talk about it . . . men have to bottle it all up until it explodes."

Interviewees' beliefs about gender and emotion, and of women as listeners were important to disclosure. Cultural display rules may dictate what emotional expression is acceptable (Levant, 1995). Six male interviewees described how their feminine side was lost in childhood. According to Kilmartin (1994, p. 13) traditional masculine socialization damages some men by rendering them "less capable of having empathic, caring, intimate relationships with other people." By the time some men seek help they may be experiencing a profound level of psychological pain. Moynihan suggests that while suffering life-threatening illness, a man may recoil "in stoical silence, desperately eager to keep hold of the masculine identity that he's been taught is symbolic of strength and success as a man" (Moynihan, 1998a, p. 13; Moynihan, 1998b). Illness disempowers. The stress experience can also be an emasculating experience for men as feelings of loss of control are magnified through the expression of painful emotion, seen as incompatible with some masculinities (Doyle, 1989).

Before reaching crisis point, all the women, but only one man, in the sample discussed their feelings with family, friends and colleagues. Within interviews men talked of emotions they were unable to disclose to others. Stephen became very upset and tearful as he told his experience. He had been unable to speak with colleagues or family about his distress – "the more things started to go wrong, the less I'm prepared to talk about it, the feelings of failure. I don't want to expose the failing too much . . . everything just seemed to add to this sense of failure."

Andrew described this new experience: "You're making me look intro-spectively. I really haven't thought about myself, how I feel, so trying to describe to someone else how I feel is very difficult." He felt "stress maybe was for wimps . . . it's not masculine to have emotions, feelings and things." Fighting his tears, he saw himself as struggling against emerging emotion rather than using it, "because I'd rather that I didn't have those feelings . . . we're all conditioned by the society we're in. I don't think society thinks that it's a masculine, grown-up thing to do. If I was a little boy, I could cry then but, you know, men don't . . . a man can show anger, a woman can burst into tears – it's maybe the same emotion that they're both displaying in different ways." Feelings of shame, failure and fear reinforced one another.

The expectation of women as providers of emotional labour was important in enabling some interviewees to disclose. Confidentiality and trust enabled both women and men to discuss problems with impotence and loss of libido, menopause and sexuality. Four men revealed they had never before spoken of these issues to anyone. Three asked for advice in helping their relationship recover. While there may have been areas where being male would have facilitated the collection of alternative data, I believe access to the feminine voice was instrumental in facilitating some areas of disclosure.

The Influence of Other Voices

The gendered voice was not the only important voice. Both academic and teacher voices were significant. Interviewees felt it important that the academic and political world take their plight seriously. They wanted to change concepts of 'failing teachers' and sought an opportunity of reaching policy makers. Andrew saw the experience as "an indirect way of helping others." Luke viewed it as a chance to "try and make sure that other people in my position can get support. I've been through it and that's why it's important, if there's anything I can do to help other people in that position, not personally, but a method to set things up." Interviewees felt the teacher voice aided under-standing. As Rebecca stated, "you understand it – the background and the pressures, and the rowdyism, the sense of responsibility you feel for the children, and the things that come up in schools, that it's become more and more like a battle ground."

The role of the counselling voice was substantial. The person-centred interview may actively encourage self-reflection in deeper ways than some other techniques. Interviews were not generally characterized by equality in power – with 14 participants being ill and nine receiving prescribed medication. As the 'healthy' individual I held considerable power over those more vulnerable. It would have been relatively easy to manipulate and to exploit, to persuade the more insecure interviewees into disclosing more information than they felt comfort in giving. Luke revealed, "After you'd gone I thought I can't believe that I've just spilt my guts out to someone that I don't know." While this caused some feelings of vulnerability, on reflection, he trusted my promise of confidentiality, and recognized that control regarding the use of his data lay with him. The degree of intimacy surprised interviewees. When I explored Charlotte's feelings about proceeding, she revealed, "I'm a bit embarrassed, I suppose . . . this is very revealing isn't it, warts and all."

It would also have been easy to accept surface emotion and move onto other areas. "Staying with the feelings" is a counselling strategy which encourages the expression of emotion, providing a deeper understanding of individual emotion through getting beyond the public face (Mackay et al., 1998). One of the primary emotions which re-emerged was anger. Luke's interview illustrates the need to distinguish between script and real emotion. He initially stated that he felt no anger, no need to express anger.

> "I don't get angry. I never get angry. I'm a very placid sort of person." There was a short pause. I raised one eyebrow while he reflected. "You're triggering things off in my mind because now I can say yes I could have felt angry at different times with my parents" – pause – "because their relationship was an important part of my life and I did get angry at times with them." He talked of anger in childhood and the need he felt then to distance himself from this negative emotion, explaining, "Oh yes, I didn't want to argue, I didn't want to shout, I didn't want to be like them." – pause – "Yes, I didn't want to be like them and in the end by not being abusive, by not giving the release, it's driven under. I don't feel the need to be angry" – pause – "Everybody does from time to time." He then explained his anger at school decisions where he'd "been mistreated and lied to . . . .. It was finding out parts of my nature I hadn't recognized before, and I did have some major angry times which took me by surprise." He realized he was 'running on adrenaline', displacing the negative emotion he felt onto performing with a band at nights, "It was just a cacophony . . . quite popular at the time but it was just a big noise. A big angry noise."

During this interview we had been interrupted several times by the cat wanting to come in and out. We were interrupted by it yet again. "That bloody cat! I'm going to get angry with it in a minute!" We laughed as he realized what he had said. Through the reflection process he recognized anger as one facet of his multiple self.

I was viewed as their 'unofficial' therapist by some. Ralph felt the research experience had been helpful, acting as "a form of therapy. It has allowed me to think through problems out loud and hopefully I've done the right thing for myself and my family." Alex saw it "as a form of counselling. You clearly haven't counselled me, you've enabled me to do what a counsellor would normally do, which is talk." As he told his story, he ensured my silence through body language. Each time I tried to enter the conversation, asking for further information or clarification, he held up an outstretched hand, palm towards me. This I initially interpreted as a gendered message of dominance. I felt a sense of powerlessness which was absent with both the women and with those men who appeared more comfortable with their 'feminine' selves. On reflection it may be that Alex required control over the process for his own self-protection. As his 'counsellor', my place in this interview was to listen.

Several voices thus contributed to the voice repertoire enabling disclosure. It was the counselling voice, however, which provided the safe environment.

Soufflé Bubbles: Managing Re-emergent Emotion

The sensitive interview can unleash feelings of distress in both interviewee and interviewer. Teachers talked of work abuse within organizational cultures of fear, of violence and sexual harassment from pupils, of relationship breakdowns and value discord. Maureen described the stress within her school as "a virus which feeds on itself. One person is unhappy, another person is unhappy. It spreads round within the community. People don't recognize how stressed they are becoming."

Teachers and family members described how health deteriorated as stress-related illness and burnout took hold. As crisis approached, they recounted problems with diabetes balance, asthma, blood pressure, depression, sleep loss, nightmares, migraine, memory loss, chronically upset stomach, teeth grinding leading to jaw dislocation, uncontrollable crying, increasing self-medication, and drug and alcohol abuse. Maureen declared, "I was screaming inside. I wasn't me, the personality just gets wiped out, the person you think of as you."

They reported feeling unvalued and devalued, out of control and worthless, bringing home feelings of emotional pain, of impotency, frustration, fear, resentment, anger, anxiety, despair, guilt, shame and failure. There were perceptions of multiple losses, in confidence, trust, and self-esteem, of power, status, identity, autonomy and respect. As perceptions of competency and control within life worlds decreased, there was a concentration on negative emotion and a loss of positive emotion. Huge emotional struggles took place within family systems, as families lost care providers, and as feelings of shock, bewilderment, anger, fear, frustration and helplessness took hold. Interviewees had all undergone severe emotional distress, several remaining traumatized by their experiences. It was vital to protect interviewees from harm and to have resources to deal with any problems that might have arisen.

Keeping Safe: The Interviewee Self
I needed safeguards both within and outside interviews. Fourteen interviewees were already in touch with counselling services/psychiatrists, had access to a general practitioner or skilled family members/friends. For others, I made contact with a counselling service. Having discussed 'opening the can' with all participants on the telephone before arranging interviews, eight had prepared themselves for possible distress, with close friends or relatives standing by. There may be a need for professional intervention between sessions. All had my telephone number and were advised to call if they needed to talk. For one of the participants, who experienced a false renewal and further distress, I suggested professional intervention might prove helpful. This was refused.

To maintain a safe holding environment within interviews, I monitored both participants' emotional state and my own. Through counselling supervision from a qualified counsellor, I learned to understand (a) transference, where the interviewee projects feelings towards significant others onto the interviewer, and (b) counter-transference, where feelings are aroused in the interviewer which provide clues to the interviewee's emotional state (Holloway, 1995; Houston, 1995; Hawkins & Shohet, 1996). While much counter-transference is unconscious, my counsellor taught me that as I became aware of feelings within the interview, I needed to ask myself two questions – Did I bring these with me? Are they new? If I recognized their origin as within me, they had no place in the interview. If I knew I did not bring them, then they needed to be noted and explored for meaning. Through such emotional labour I became more skilful in using the counselling voice in recognizing, assessing, monitoring and managing emotion within interviews.

Young & Lee (1996) argue that "the emotion work that is done in research is best seen as an attempt to manage feeling rules" where "dissonance is not only inevitable but offers potentially valuable insights into the competing tensions of involvement, comfort, and identification" (in James & Gabe, 1996, p. 16).

Emergent thoughts provoked dissonance. As potentially disabling feelings bubbled to the surface, I gave continual reassurance, supporting as new and old thoughts and feelings emerged. Marcus talked of feelings of guilt towards colleagues providing cover while he was absent. I reflected back his words, asking whether he had similar feelings of guilt towards pupils. He replied, "I didn't feel guilty because the pupils were missing out if that's what you mean. No, I never did actually, perhaps I should have done!" He laughed and declared "Now look what you've done! No, no I didn't!" We both laughed. I gave him time to think. He realized he felt differently to some colleagues, "I was angry probably with the pupils . . . it was the pupils who got me into that situation. In the end I just said, sod it! I don't care any more. I got to that stage, which is very unlike me, not to care. So it changed my outlook. I just turned off." Laughter dissipated the tension of dissonance. Time to reflect achieved consonance.

'Feeling rules' were not only managed, they were also transgressed. Luke has many unpleasant memories which he found difficult to express. He was very tearful throughout his first interview. During tearful episodes, I gently supported him until he felt composed. Counselling training has equipped me to feel comfortable with tears and with silence, giving time for self-reflexivity, enabling interviewees to gather thoughts and find emotional expression. A symbiotic relationship developed. As Luke's voice lowered when talking of the situations which led up to breakdown, the bizarre behaviour, hallucinations and fears, my voice lowered to match. He was given psychotropic drug therapy,

"I was awake but completely comatose, then my body started to come back, but it was uncontrollable . . . . I could move about but it also moved itself. It was like having Tourettes. I was very scared.' There was a long pause. When he signalled readiness to continue through re-establishing eye contact, I repeated the last phrase back to him. He continued his story with many long pauses and tears. Apologizing, he explained his motivation to continue, "Sorry. It's quite difficult emotionally . . . I don't like feeling out of control, that's really what I was – out of control or no control, running around doing crazy things. I feel a bit better now. This is a bit of a monologue. I'm OK . . . I want to make sure that other people in my position can get support. I've been through it. That's why I think it is important that I do this to help other people . . . that was quite painful. I don't go back that often. I do at times, and I think it's necessary. It comes back to being part of this process of being better that you can recognize the symptoms that take you back to what started it off."

Negative emotions, past and present, were worked through, explored only when participants felt ready to do so. Through acknowledging and reflecting feelings, the feeling rules within the interview experience were signalled. The expression of painful emotion, such as anger, sadness, grief, despair, often discounted as irrational and forbidden, was deemed appropriate and 'norm-alised'. It was alright to cry. It was alright to show anger.

Some emotion, however, may best be left to experts. Rebecca explained her 'feeling rules' for keeping herself safe, saying, "I couldn't cry for years. I was afraid if I cried or let any emotions out, I suppose I had so much to cover, that it would come out like an explosion and I wouldn't be able to get the lid on and I wouldn't be able to function. So I was petrified." A disturbing assault by a pupil resulted in post-traumatic stress disorder and emotion from a childhood experience resurfacing. She explained "I've been totally out of control for over a year. . . . I couldn't recognize myself . . .. I had a series of six sessions with [counsellor], to talk it through, talk about what it had triggered in me, strange thoughts and feelings."

Interpreting her body language and my feelings as indicating unreadiness, I did not pursue this and changed direction. In a later session she brought up this incident again. She had recently arranged sessions with a psychotherapist.

> D: You said that this incident brought up all sorts of feelings and anxieties?

> R: A similar situation in childhood – I'll perhaps talk about them later.

Rebecca never did explore these emotions with me. I could have probed further but did not. I had no right to ask her to open her can against her will. She knew what might be below the surface and did not feel ready to disclose them here. Rebecca's own capacity for self-determination ensured her 'can' remained closed and in the hands of her chosen expert. I asked myself how often, within longitudinal interviews, I had the right to ask interviewees to open their can.

Feedback was one area where the resurfacing of emotion re-opened the can. I offered to provide brief summaries of interviews. This created an added stressor for some. Maureen, a participant in the pilot study, found reading her narrative summary a painful experience, both disturbing and illuminating. After interview five, she read her summary alone. By the time she finished she was crying. She initially felt anger against me. Seeing her 'special places', locations which held special spiritual meaning for her, in black and white was distressing. She did not want others to read about them. Some terminology – e.g. the 'hello/good-bye' effect – she did not understand. Her husband was very upset at the initial effect this had on her, interpreting her response as rekindling distress. While initially disturbing, it made her think very hard about her feelings. She had recently heard of several colleagues now absent diagnosed as suffering stress-related illness. She realized just how angry she felt. She had adopted an individual stress discourse, directing anger at herself for being a 'non-coper'. She now adopted an organizational stress discourse, directing her feelings towards an abusive system. Insensitive handling of feedback could have lost a valuable conversational partner. I was able to reassure her. Discussion before interview six indicated Maureen viewed emotional pain as necessary to self-renewal. She interpreted her distress as helping effect closure on the past and maintained her motivation to continue with the study.

Maureen's experience taught a valuable lesson. I had underestimated the impact of the narrative summary – both verbal and written. I had agreed to bring Luke the first interview summary. Before turning on the tape, over a cup of tea, we discussed his thoughts during the intervening month, then read through the summary together. We worked through some of the emotion which surfaced. He was close to tears. When he felt he could talk on tape the conversation began:

D: How did you feel reading that?

L: It's quite painful.

D: Quite painful?

L: I've not been through the whole process myself and written it all down. As I was reading, I was visualizing events that we discussed. I think when I visualize the event I get the same emotions through my head as at that time. They're still there now and I think that's why it's painful. I think some of the things are quite shocking particularly the bit where my life is running really out of control. The part about the drinking. Some of this behaviour was quite self-destructive, but I can see that now, as I'm reading that, that I'm quite a different person in lots of ways, and not in some ways, that it can still be part of me, still part of my makeup, but I've very much moved on in my experiences

Beginnings and endings of interviews need careful management. Before ending interviews I spent time making sure interviewees were feeling comfortable, bringing them back to the 'real' world, often over coffee, taking the conversation into other areas of life. Luke shared his feelings with me over a cup of tea. He had found it very difficult to talk about emotion. Before self-renewal, emotion had indicated negativity. He felt he had no vocabulary for feelings in words, more in sound, colour, images. "I've got no words to describe it so I draw music around it." He played me some of the music he was working on. His very intense tone poems of emotion touched me deeply. Here, I too felt total inadequacy in verbal expression. We parted friends. I was invited back during the summer to visit him and his partner. I wrote in my diary "[His music] reminded me of Fred Hoyle's novel *The Black Hole*, reaching deep into the soul, going on this journey and emerging as part of the expansive oneness with the universe, a feeling of freedom, of being at one, a merging of wisdom, like dissolving but not into non-existence – into pure existence." Perhaps his metaphor for self-renewal? To be able to provide a secure environment for interviewees meant I needed to feel secure myself. I found I also needed a safety net.

Keeping Safe: The Fieldworker Self
I had to manage my own emotion. "You must be soaking up all their emotion. How do you cope with it?" demanded an ex-colleague. Recognizing and

managing my own emotion state involved noting my feelings but remaining silent, which I found difficult at times! There were remarks made about women, especially where relationships had ended, which I found uncomfortable. I felt anger at the way some treated women in their lives, their feelings about the feminization of their profession, towards women in positions of power, the language used and why I sometimes seemed invisible as a woman. At times I found listening to this offensive and distressing. I experienced a loss of power. There was conflict between what I felt I 'ought' to be feeling (unconditional positive regard, empathy, etc.), what I was feeling, what I wanted to feel, and what I tried to feel, emotive dissonance giving clues to my feeling rules. A colleague engaged on similar research with primary school teachers avoided getting into intense emotional situations. He found that during tearful episodes he experienced 'acute embarrassment' (Troman, 1998). While I am used to tears and outbursts, both my own and those of children, husband, friends, pupils, colleagues, I felt acutely uncomfortable with the expression of emotion on two occasions.

According to Deborah Lamb, "All research is me-research' in that it may entail confronting one's own 'ghosts'" (Ely et al. 1991, p. 124). One of the most useful therapy techniques has been the 'muck-cart'. I was once told to accept all the projected feelings thrown at me in a counselling encounter, to store them temporarily on my shoulders, but as I left to return home, to heave them on a muck-cart and leave them behind. After these interviews I was able to walk away from participants and metaphorically throw those feelings into that cart. I stopped in a lay-by to make field notes. By the time I was home, negative emotion had usually been worked through. I was successful in all but two interviews, one of which completely overwhelmed me as it touched my own submerged emotional past. Analytical distancing became temporarily impossible as my own can opened.

This possibility had been anticipated within the research design. I was funded for six sessions with a counsellor of my choice. My colleague Geoff was not given the same provision. He now wonders if this might have been useful for him since it was "hard to listen to bad news all day" (Troman, 1999). In many professions, such supervision is the norm on a much more regular basis. Six sessions were not enough. It did not cover the full span of interviews. Unfortunately, counselling provision had ceased when my own can opened. I was lucky to have family and friends who supported me. The resources of a counsellor experienced in work counselling supervision proved invaluable (Carroll, 1996). The sessions provided 'coping' tools: butterflies on my fridge and computer to remind me of how she carefully brought me out of my chrysalis, a 'parrot' on my shoulder which repeatedly told me I was OK. They contributed to the research through allowing me to explore interview resistances (my reluctance and interviewee reluctance to pursue certain areas), transference and counter-transference issues and paralleling (where processes at work within the interviewee/interviewer relationship are mirrored in the interviewer/counsellor relationship). They aided my own personal growth. Through the experiencing of vicarious emotion, I gained greater understanding of the difference between the emotional empathic voice – the experiencing of other's emotion, and the intellectual empathic voice – taking the perspective of the other (Duan & Hill, 1996). This enhanced the empathic voice. The research experience became a positive one for both interviewees and interviewer.

THE RISEN SOUFFLÉ: THERAPEUTIC OUTCOMES

In therapeutic work, it is expected that there are therapeutic outcomes. In sensitive research, there may also be therapeutic outcomes. For some interviewees, there were profound changes in understanding their emotions.

Charlotte and I did much exploration of both negative and positive emotion within her last few years in teaching. A new understanding of values emerged. "It's very important. I wanted to carry on, give myself a quality life, one that I was proud of, because the more it was going down the pan, the more I felt I was going back into using drugs again. But I didn't realize until now, until talking to you, I didn't realize. Isn't that strange?"

I wrote in my diary, "I often feel I am exploiting them." This was dismissed by participants. Checking this out, I found no one admitted to feeling exploited. When I asked Charlotte she said no and explained how beneficial it had been for her.

> Until talking to you I didn't realize. It's really a nice feeling, to feel that I'm not on my own. I think something like stress, you shelve it under the carpet but to be able to actually pick it up and look at it for what it is, it's an enrichment in life really. It's been good, made it three-dimensional for me. Exploring it has made it tangible and by making it tangible, you can pick it up and plonk it away. It's actually made me look around it, at the nooks and crannies of it, walk around it, and it's not such a horrible thing as it was. It's not very big at all – a wrinkle in life! You've given it a label so you can see it as a symbolic thing, it's like this magic thing, you've named it and so you can now get rid of it, exorcise it because you've named it and given the devil it's name, Beelzebub and the devil is out!

This experience was cathartic for Charlotte. It enabled her to objectify the stress experience and the self, and helped her come to terms with some losses. Marcus revealed,

> To begin with, I think it made me think about the situation and analyze it a little bit more, and then the last couple of sessions, it's been rather strange in that I feel divorced now from

the person who was stressed out three months, or however long it was ago, I feel that person doesn't exist any more, because it's all gone, it's all in the past now, it's something which is almost not me if you like, it's a part of my life which I've shelved now and I've gone on from it.

The interviews were instrumental in his being able to accept closure of this period of his past. I asked Rebecca, ". . . one of the fears I had was in opening up the 'can of worms'. I wondered how you felt about this?" She replied:

At first I was very fragile wasn't I? I haven't spoken in such depth about it. I feel it's been really interesting. It's made me look very closely at what happened and I think the fact that we've given each other this month apart and isolated the stages has been immensely interesting you know like unresolved and resolved, and I think some of the comments that you've made and some of the bits you've taken out of it and shown to me back are really interesting, to clarify it and I've found it very beneficial, I have, extremely, extremely.

There was a great deal of pain to work through but it was seen as essential to the healing process. As Rebecca explained:

I think when I started looking inwards at some of my own problems, I found that difficult, but not negative, it had got to be done, got to be done. I found that hard. I'm the sort of person if there's something there I'd rather do it, I'd rather take it out and look at it and deal with it . . . A couple of times early on I remember I felt tearful but I think it was a healthy tearful because it was dealing properly with something.

Some months later we had a cup of tea and a chat in her new workplace. I knew she was fine.

Ralph found that interviews aided earlier recognition of an illness resurgence after false renewal, "it's been helpful – I don't think it's made me think about anything I haven't wanted to think about . . . I suppose that's the success of the conversations I've had with you and [his female counsellor], made me put me at the centre of things." Changes in feeling and thought throughout narratives were apparent. As far as I am aware, the benefits outweighed any losses.

In later interviews, I sought insight into how interviews had affected participants, allowing me to gauge readiness to stop contact. Alex declared "It's been a pleasure." Gareth had very much enjoyed the experience, looking forward to receiving analytical feedback. Participants overwhelmingly found interviews positive experiences, exploration in safety providing new under-standings. I interviewed Jonathon's partner Jo. Jonathon felt it helped their relationship, saying, "I think it gave her a better insight into some of the events and things that had gone on between us so perhaps – it should happen to everybody, you know, you shouldn't just counsel the one part of the family, you should counsel both parts of the family."

Luke felt his interviews aided his relationship with his partner. "I felt a big release having talked to you . . . some of those memories that I've discussed with you at a personal level, with my partner whom I've known for X years,

I've not even discussed them with her . . . last Friday night, lots of that came out and we had a big cleansing between us, lots of things that she got out, it all came out and that's something that I probably wouldn't have done before. I recognize that as a step forward."

Only one interview was observed by a colleague. As Harry explained how his life had been turned upside down by his partner's experience of stress-related illness, he wept, hiding his face behind his arm, tearing at his hair. One of his main losses was the warmth, the closeness, the bodily contact in his relationship with his partner. At the close of the interview, he came forward and shook my hand. My nurturing voice asked if he would accept a hug. We held each other briefly. He kissed me lightly on the cheek. This is perhaps not a response many male fieldworkers would feel comfortable with. Perhaps had I been alone I might have felt this unsafe. However, I feel I gave Harry something he needed at that time.

I found that I not only gave to interviewees. They also gave to me. Interviewees, particularly females, were attentive to my emotional needs through seeking mutual disclosure and offering support. However, it was Luke who created a safe post-interview holding environment for me, encouraging me to open up to him. We had developed an intense rapport. He had picked up some counter-transference from me. While we finished the session with another cup of tea, he asked how I was feeling, sensing I was troubled – perhaps I might feel free to talk with him in the role of 'professional stranger'? I was hesitant. After two and a half hours of intense conversation, my 'mask' was down, my non-verbal communication showing he had hit the proverbial nail on the head. He again repeated his offer. I ended up spending half an hour 'spilling my guts'. After this I experienced what I must assume to be similar emotion to his regarding trust and confidentiality issues. While I had promised confidentiality to him, he was not under the same professional commitment to me. I felt power relations had been reversed. I felt extremely vulnerable but I also felt trust in him. Luke was well in tune with both masculine and feminine gender orientations. I felt in tune with his voice repertoire. This experience with Luke showed that it is not being a woman *per se* that is of prime significance in the sensitive interview relationship, but possession of a flexible gender role orientation, feeling comfort with both feminine and masculine aspects of the self, access to particular voices within particular social contexts, the capacity to perform the necessary emotional labouring, and the 'fit' between individuals.

Last words
When I began this research, I thought I was well-prepared. The original focus was self-renewal after stress-related illness and burnout. However, I had to

broaden my skills base as the full impact of the emotion aspects of stress-related illness became clearer. A few years ago I would not have reached a successful outcome. Life has provided an immensely fruitful learning curve, where the skills of intellectual empathy, empathic emotion and empathic listening skills have been sharpened through a variety of encounters which provided valuable opportunities for practice. Through PhD supervision and work counselling supervision, through my own personal growth and through the voices of interviewees, my voice repertoire expanded, enhancing emotional literacy, and enabling listening for many interviewee voices rather than few.

To explore emotion requires a particular mix of 'skills' such as the ability to self-disclose, to demonstrate vulnerability, to share power and intimacy. It requires considerable emotion management skills and emotional literacy. Many women learn these skills within their relationships with others. Not all women develop these emotional skills. Much work on gender claims that men generally are less likely to possess these skills or to feel comfortable with emotion expression. However, there are many men whose voice repertoires provide them with the necessary emotional skills.

I view the success of this project as a result of being positioned both as a female and as a counsellor, and to the creation of an appropriate skills base. For some interviewees, especially those men holding rigid gender role orientations, it undoubtedly was important that I was female, that I did not compound any loss of face, that I did not threaten loss of identity further, and that I shared control of the interview process. Of most significance, however, was the combination of voices which produced an outcome which, for the majority of interviewees, was thought provoking, enlightening, self-enhancing and growth promoting, contextualizing experience and helping them get on with the rest of their lives.

It is not only when researching vulnerable populations that emotion comes to the forefront. Emotion lies at the heart of much ethnography. Participants have complex lives, complex histories. Organizations have complex emotion environments. Researchers have no crystal ball to see what questioning might provoke. In carrying out interviews fieldworkers may enter personal and organizational worlds, open the 'can of worms', follow the recipe, but not always be there to see what emerges from the oven.

Opening the can may be a tricky business. Appropriate tools may not be available. The 'worms' may go off in unexpected directions. Baking the soufflé may release unforeseen turbulence. The 'can of worms' and the 'worm soufflé' are thus important concepts. Do all fieldworkers engaged in such work possess skills which enable them to explore the 'worms' effectively? What opportunities are available to hone these skills through professional development? In raising consciousnesses, what duty of care does the research community hold towards interviewees and towards field workers? What ethical guidelines are available? What safeguards are provided? What 'can of worms' might the research community hold? It is time to give gender and emotion a higher priority in the professional development of research students and field-workers.

REFERENCES

Abrams J., & Zweig C. (Eds), (1991). *Meeting the Shadow: The Hidden Power of the Dark Side of Human Nature*. New York: Putmans.
Arendell, T. (1997). Reflections on the Researcher-Researched Relationship: A Woman Interviewing Men. *Qualitative Sociology, 20*(3), 341–368.
Bown, G. (1998). Seeing, Feeling and Working Differently. Paper presented at Healthier Organisation Seminar, Stress at Work, Northampton.
Brannen, J. M. (1988). Research note: The Study of Sensitive Subjects: Notes on Interviewing. *The Sociological Review, 36*(3), 552–563.
Bulan H. F., Erickson, R. J., & Wharton A. S. (1997). Doing for Others on the Job: The Affective Requirements of Service Work, Gender, and Emotional Well-being. *Social Problems, 44*(2), 235–256.
Cannon, S. (1989). Social Research in Stressful Settings: Difficulties for the Sociologist Studying the Treatment of Breast Cancer. *Sociology of Health and Illness, 11*(1), 62–77.
Carroll, M. (1996). *Workplace Counselling: A Systematic Approach to Employee Care*. London: Sage.
Cowles, K. V. (1988). Issues in Qualitative Research on Sensitive Topics. *Western Journal of Nursing Research, 10*(2), 163–179.
Davies, B., & Harré R. (1990). Positioning: The Discursive Production of Selves. *Journal for the Theory of Social Behaviour, 20*(1), 43–63.
de Beauvoir, S. (1949). *The Second Sex*. London: Penguin.
Doyle, J. A. (1989) *The Male Experience*. Iowa: Wm C. Brown.
Duan, C., & Hill, C. E. (1996). The Current State of Empathy Research. *Journal of Counselling Psychology, 43*(3), 261–274.
Eichenbaum, L., & Orbach, S. (1983). *What Do Women Want: Exploding the Myth of Dependency*. London: Coward-McCann.
Ellis, C., & Flaherty, M. G. (1992). *Investigating Subjectivity: Research on Lived Experience*. California: Sage.
Ely, M. Anzul, M., Friedman, T., Garner, D., & Steinmetz McCormack, A. (1991). *Doing Qualitative Research: Circles within Circles*. London: Falmer Press.
Francis, E. (1997). Developing Philosophical Voice. Paper presented at Symposium on Philosophy and Educational Research, Bera, York.
Gagnon, J. H. (1992). The Self, its Voices and their Discord. In: C. Ellis & M. G. Flaherty (Eds), *Investigating Subjectivity: Research on Lived Experience*. Sage: California.
Goleman, D. (1995). *Emotional Intelligence*. New York: Bantam Books.
Gornick, V. (1971) Woman as Outsider. In: V. Gornick & B. K. Moran, (Eds), *Woman in Sexist Society*. New York: Basic Books.

Graham, P. A. (1995). Teacher Burnout. Paper presented at conference on Teacher Burnout, Marbach, November.

Hawkins, P., & Shohet, R. (1989). *Supervision in the Helping Professions*. Buckingham, Open University Press.

Hearnshaw, H., Reddish, S., Carlyle, D., Baker, R., & Robertson, N. (1998). Introducing a Quality Improvement Programme to Primary Health Care Teams. *Quality in Health Care, 7,* 200–208

Hochschild, A. R. (1983). *The Managed Heart*. Berkeley: University of California Press.

Hochschild, A. R. (1997). *The Time Bind: When Work becomes Home and Home becomes Work*. New York: Metroplolitan Books.

Holloway, E. L. (1995). *Clinical Supervision: A Systems Approach*. California: Sage.

Houston, G. (1995). *Supervision and Counselling*. London: The Rochester Foundation.

James, N. (1989). Emotional Labour: Skill and Work in the Social Regulation of Feelings. *The Sociological Review*, 15–42.

James, V., & Gabem J. (Eds), (1996). *Health and the Sociology of Emotions*. Oxford: Blackwell.

Karniol, R., Gabay R., Ochion Y., & Harari Y. (1998). Is Gender or Gender-role Orientation a Better Predictor of Empathy in Adolescence? *Sex Roles, 39*(1/2), 45–59.

Kilmartin, C. T. (1994). *The Masculine Self*. New York: Macmillan.

Lather, P. (1986). Research as Praxis. *Harvard Educational Review, 56*(3), 257–277.

Lee, R. (1993). *Doing Research on Sensitive Topics*. London: Sage.

Levant, R. F. (1995). Towards a Reconstruction of Masculinity. In: R.F.Levant & W.S.Pollack (Eds), *The New Psychology of Men*. New York: Basic Books.

Mac an Ghaill, M. (Ed.) (1996). *Understanding Masculinities: Social Relations and Cultural Arenas*. Buckingham: Open University Press.

Mackay, H. C., Barkham, M., & Stiles, W. B. (1998). Staying With the Feeling: An Anger Event in Psychodynamic-Interpersonal Therapy. *Journal of Counselling Psychology, 45*(3), 279–289.

Mearns, D., & Thorne, B. (1988). *Person-centred Counselling in Action*. London: Sage.

Miller, A. (1991). *Breaking Down the Wall of Silence*. London: Virago Press.

Moynihan, C. (1998a). Strength in Silence. *The Guardian*, October 20.

Moynihan, C. (1998b). Theories of Masculinity: Theories in Health Care and Research. *British Medical Journal, 317,* 1072–1075.

O'Brien, M. (1994). The Managed Heart Revisited: Health and Social Control. *The Sociological Review*: 393–413.

Oakley, A. (1981). Interviewing Women: A Contradiction in Terms. In: H. Roberts (Ed.), *Doing Feminist Research*. London: Routledge and Kegan Paul.

Paterson, A. S. (1997). A Humanistic Framework for Interviewer Skills. Paper presented at Student BERA, York.

Pearlin, L. I. (1989). The Sociological Study of Stress. *Journal of Health and Social Behaviour, 30,* 241–256.

Rogers, A. G., Brown, L. M., & Tappan, M. B. (1994). Interpreting Loss in Ego Development in Girls: Regression or Resistance? In: D. Lieblich & C. Josselson (Eds), *The Narrative Study of Lives: Exploring Identity and Gender*. California: Sage.

Rogers, C. R. (1951). *Client-centred Therapy: Its Current Practice, Implications and Theory*. Boston: Houghton Mifflin.

Schaufeli, W. B., Maslach, C., & Marek, T. (Eds) (1993). *Professional Burnout: Recent Developments in Theory and Research*. Washington: Taylor and Francis.

Sharpe, S. (1976). *'Just Like a Girl': How Girls Learn to be Women*. Harmondsworth: Penguin.

Stacey, J. (1991). Can There Be a Feminist Ethnography? In: S. Gluck & D. Patai (Eds), *Women's Words: The Feminist Practice of Oral History*. London: Routledge.

Stanley, J. (1993). Sex and the Quiet Schoolgirl. In: P. Woods & M. Hammersley (Eds), *Gender and Ethnicity in Schools*. London: Routledge.

Stanworth, M. (1983). *Gender and Schooling: A Study of Sexual Divisions in the Classroom*. London: Hutchinson.

Stapley, L. (1996). *The Personality of the Organisation: A Psychodynamic Explanation of Culture and Change*. London: Free Association Books.

Statham, J. (1986). *Daughters and Sons: Experiences of Non-sexist Childraising*. Oxford: Blackwell.

Strauss, A. L. (1987). *Qualitative Analysis for Social Scientists*. New York: Cambridge University Press.

Tannen, D. (1991). *You Just Don't Understand: Women and Men in Conversation*. London: Quality Paperbacks Direct.

Tolich, M. B. (1993). Alienating and Liberating Emotions at Work: Supermarket Clerks' Performance of Customer Service. *Journal of Contemporary Ethnography*, 22(3), 361–381.

Troman, G. (1998). *Does Gender Make a Difference? A Male Researcher's Reflexive Account of Gendered Fieldwork Relations in Ethnographic Work on Stress in Teaching*. Paper presented at Ethnography and Education Conference, Oxford, September.

Troman, G., (1999). Personal communication.

Warren, C. A. B. (1988). *Gender Issues in Field Research*. California: Sage Publications.

Williams, S. J., & Bendelow, G. (1996). Emotion, Health and Illness: The 'Missing Link' in Medical Sociology? In: V. James & J. Gabe (Eds), *Health and the Sociology of Emotions*. Oxford: Blackwell.

Woods, P. (1982). Conversations with Teachers: Some Aspects of Life-history Method. *British Educational Research Journal*, 11(1), 13–26.

Woods, P. (1986). *Inside Schools: Ethnography in Educational Research*. London: Routledge.

Woods, P. (1995). Intensification and Stress in Teaching. Paper presented at conference on Teacher Burnout, Marbach, November.

Young, E. H., & Lee, R. (1996). Fieldworker Feelings as Data: 'Emotion Work' and 'Feeling Rules in First Person Accounts of Sociological Fieldwork'. In: V. James & J. Gabe (Eds), *Health and the Sociology of Emotions*. Oxford: Blackwell.

DOES GENDER MAKE A DIFFERENCE? A MALE RESEARCHER'S REFLEXIVE ACCOUNT OF GENDERED FIELDWORK RELATIONS IN ETHNOGRAPHIC WORK ON STRESS IN TEACHING

Gender is a focal, an organizing category in social life and social science. What this means for field researchers is that we must develop a sensitivity to issues of gender both when we do our fieldwork, and when we write it up. Since our methodology has as its central commitment the unity of the interpretive process – of the biography and history of the researcher – then it is vital to understand the place of gender in social research (Carol Warren, 1988, p. 7).

INTRODUCTION

Those researchers working in the anthropological tradition of ethnography acknowledge that the main instrument in this form of research is the researcher's self (Burgess, 1984; Walford, 1991). Recent methodological developments (the growth in ethnographic style research, life-history and narrative methods) in this field have shown the importance of researchers adopting a reflexive attitude in order to understand how the interaction between the self and other influences both the research process and researchers' accounts of the social world. It is the personal and social characteristics (age,

Genders and Sexualities in Educational Ethnography, Volume 3, pages 209–230.
Copyright © 2000 by Elsevier Science Inc.
All rights of reproduction in any form reserved.
ISBN: 0-7623-0738-2

social class, sex, gender, race/ethnicity) of the researcher which shape the research process. With the exception of gender each of these characteristics has been given a good deal of attention in the methodological literature written by men (Hammersley & Atkinson, 1995). However, sex and gender issues in fieldwork have tended to remain the focus of feminist scholarship, thus reflecting the postmodern concerns with reflexivity in the 'fifth moment' of ethnographic research (Delamont, 1997; Denzin, 1997). In anthropology, however, gendered reflexivity is not new. Since the early part of this century cultural anthropology has often been conducted by wife-husband teams (see for example Friedl (1980) on researching Greek culture; Freedman (1986) on Romania; Fleuhr-Lobban & Lobban (1986) on the Sudan). Quite apart from married couples gaining credibility and 'respectability' in many cultures, the pair are better equipped to explore cultures which are gender segregated. Female anthropologists have typically confined themselves to "women's and children's worlds" (Warren, 1988, p. 16). Their research has focused on "'women's issues' and women's settings, mainly the domestic sphere of child rearing, health and nutrition" (p. 16). Sudarkasa (1986, p. 181), for instance, records in her study of the Yoruba:

> I was never expected to enter into, and never did see, certain aspects of the life of men in the town. I never witnessed any ceremonies that were barred to women. Whenever I visited compounds I sat with the women while the men gathered in the parlors or in front of the compounds . . . I never entered any of the places where men sat around to drink beer or palm wine and to chat (cited in Warren, 1988, p. 16).

In addition to their gaining access to settings denied to male anthropologists female fieldworkers are also held to be superior in their work because their "femaleness' endows them with 'the ability to communicate and gain 'confessional rapport'." (p. 42). When working with women and men the female fieldworker encounters "more willingness . . . to allow access to inner worlds of feeling and thought." Female fieldworkers seem at once less threatening and more open to emotional communication than men (p. 42). Nader (1986, p. 114) argues that:

> Women make a success of field work because women are more person-oriented; it is also said that participant observation is more consonant with the traditional role of women. Like many folk explanations there is perhaps some truth in the idea that women, at least in Western culture, are better able to relate to people than men are (cited in Warren, 1988, p. 42).

Examples of gender impacting on research are many. Warren (1988) reports that in a study of a nudist beach:

> Women and men team members found that people on the nude beach told them different things. Single men told Rasmussen about their sexual interests, but provided the woman

researcher (Flanagan) only with the rhetoric of 'freedom and naturalism'. The reverse was true when Rasmussen interviewed women and Flanagan interviewed men (Warren, 1988, p. 31).

The strengths of the female fieldworker are clear when we consider the difficulty/impossibility of a man researching such topics as: the experiences of motherhood (Oakley, 1979); the stepfamily (Hughes, 1992); Patahn mothers (Currer, 1992); female marital difficulties (Brannen & Collard, 1982); women with breast cancer (Cannon, 1989); women's experiences as AIDS/HIV sufferers (Lather & Smithies, 1997); the wives of naval personnel (Chandler, 1990); clergymen's wives (Finch, 1984); friendships in adolescent girls (Hey, 1997) and the childhood worlds of very young children (Boyle, 1996).

There are advantages, though, in adopting a marginal role. It does seem to be necessary to prevent feeling 'entirely at home':

> from the perspective of the 'marginal' reflexive ethnographer, there can thus be no question of 'total commitment, surrender' or 'becoming'. There must always remain some part held back, some critical and intellectual 'distance'. For it is in the space created by this distance that the analytical work of the ethnographer gets done (Hammersley & Atkinson, 1995, p. 115).

Marginality has other strengths. Woods (1996), for instance, invokes Aoki (1983) who, in "commenting on his probing for the essence of what it means to be human," remarks that:

> This kind of opportunity for probing does not come easily to a person flowing within the mainstream. It comes more readily to one who lives at the margin . . . It is, I believe, a condition that makes possible deeper understandings of human acts that can transform both self and world, not in an instrumental way, but in a human way (cited in Woods, 1996, p. 151).

However, the advantages of women researching women have been argued for some time. Oakley (1981) has perhaps been particularly influential. In her article advocating females interviewing females she critiques male survey researchers interviewing women. Elements of her critique are as follows:

• In striving for the 'proper' interview men seek to achieve both rapport and detachment (a traditional feature of qualitative work) with women interviewees in order to gain information but remain 'objective'. Oakley sees this as pseudo-rapport and as manipulative. Men see getting involved with interviewees as 'bad'.

• Men are instructed, by methodological texts on survey research written by men, to avoid answering the questions of those they interview in order to stay in control of the interview process.

- Women are constructed as 'passive' respondents. This 'objectifies' women and is true of both the 'mechanical' survey research interviews and pyschoanalytic/pyschotherapeutic interviews.[1]
- Male methodologists construct men as cognitive, intellectual and rational whereas women are sentimental or emotional.
- There is an asymmetry of power in the male (interviewer) female (interviewee) relationship. It is hierarchical with the male interviewer enjoying the most power. This means that not only do men get low quality data but also that women are vulnerable to exploitation.
- Men have no shared experiences of being female and therefore lack insights into women's worlds. Men, usually, cannot become close friends with the women they interview.

Oakley decided to abandon the masculinist 'text-book code of ethics' (p. 48) not simply to do better sociology but to give greater visibility "to the subjective situation of women in sociology and society." She also sought to use the feminist interview as a "tool for making possible the articulated and recorded commentary of women on the very personal business of being female in a patriarchal capitalist society" (p. 49).

Oakley's early writing has made an enormous impact on not only feminist theory and methodology but also the way researchers (women and men) view subjectivity in research. The work of Ellis & Flaherty (1992), for example, is strongly influenced by feminist subjective approaches but argues for the inclusion of both the subjective and scientific view in sociological research. The "goal is to arrive at an understanding of lived experience that is both rigorous – based on systematic observation – and imaginative – based on expressive insight" (Ellis & Flaherty, 1992, p. 5, cited in Woods, 1996, p. 107).

Finch is more general in her attack on masculinist methodology. She argues that:

> ... the 'ethics' of research are commonly conducted within a framework which is drawn from the public domain of men, and which I find at best unhelpful in relation to research with women' (1984, p. 71).

Finch abandons the traditional criterion of objectivity (a masculinist concept) and distance because "both parties (female interviewer and female interviewee) share a subordinate structural position by virtue of their gender" (p. 76). She is "startled by the readiness with which women talked to her" (p. 72). She was in a state of [shared powerlessness with women because of their structural position" (p. 86). The interviewer/interviewee relationship she had in her research would not have been possible if she had tried to maintain an "unbiased and objective distance from the interviewees" (p. 74). Finch recognizes the vulnerability of women respondents to exploitation in research and shares Oakley's political commitment as a feminist researcher.

Having read most of this literature at an early stage in the research project, it was with some apprehension that I embarked on fieldwork that was to involve in-depth longitudinal interviews with a sample of teachers (composed of at least 50 percent women) on the sensitive issue (Lee, 1993) of stress in teaching (ESRC – 1997–2000, R000237166). It is the aim of the research to have conversations with a number of teachers who have recently experienced stress (see below for sample details). We are interested in their perceptions of the causes of stress, the experience and effects of stress, their coping strategies and processes of recovery and self-renewal. Given that all the respondents have recently experienced stress and protracted illness (some experiencing clinical depression), the research interviews are potentially highly-charged emotional events. Even though I had read critiques of the feminist position on access, field relations, interviews, analysis and representation (see for example Chandler, 1990), to try to allay feelings of inadequacy in face of the task, I could not rid myself of the idea that a female researcher might be more suited for this style of research and nature of topic. This chapter, therefore, is a reflexive account of my experiences of interviewing and field relations in the stress project. I compare my experiences of interviewing men and women and locate my discussion in the methodological literature on gendered research, where male accounts of these aspects of the research process are largely absent as far as I am aware.

Researchers working in the interactionist ethnographic tradition acknowledge the salience of the researcher's self as the prime research instrument. Recent methodological developments in this field have shown the importance of researchers adopting a reflexive attitude in order to understand the ways in which interaction between the self and other shapes both the research process and accounts of the social world. While a number of social characteristics have been taken into account in this process, reflections on the impact of the researcher's gender on fieldwork and interpretation have been almost wholly produced by female researchers engaged in feminist ethnography. Writers of 'we-women' accounts have shown the methodological strengths when female researchers research women. By way of contrast, this chapter seeks to explore the impact of the male researcher's gender on field relations and on in-depth interviewing. The main concerns here are how the researcher manages the tensions between involvement/subjectivity and distance/objectivity. Additionally, the gendered responses of the respondents to the research process are described. The chapter concludes by arguing that while the researcher's self is central in ethnographic research gender is only one aspect of that self.

Before doing this I provide essential details of the research methods.

METHODS

In a research project immediately prior to this current research on stress, I was interested in the restructuring of primary schools and the impact this was having on primary teachers' work (Troman, 1997). This research involved two years fieldwork in a primary school and the subsequent writing of an ethnographic account. In this large primary school, gender was a significant factor in the composition of the staff (my research sample). Of 18 teachers, three were male and 15 female. Of the three male teachers, one was the headteacher and another the deputy head. I left this research with the feeling that I had not fully engaged with some of the issues involved in a man researching a largely female workplace (albeit patriarchally dominated) and attempting a representation of female work and experience. These issues were a recurrent feature of meetings of the research team of which I was a member at the time of the research. A colleague in this group frequently raised questions (Jeffrey, 1998) about his own research in primary schools and the dilemmas and tensions involved when men research and attempt the representation of women's lives. This situation was partly being resolved by his use of discourse theory (Davies, 1992) in his interpretation of the data.

My main research methods at this time were participant observation and 'informal' interviewing. The majority of data were collected in observation and informal conversation; very little 'formal' interviewing took place. By contrast, the principal research method of the stress research is semi-structured, open-ended, in-depth, life history interviewing. It was a matter of priority, therefore, for me to familiarize myself with the literature on the interview methodology and initially I consulted the classic methodological texts (Burgess, 1988; Finch, 1984; Hammersley & Atkinson, 1995; Oakley, 1981; Woods, 1986).

Most research on teacher stress has adopted a psychological perspective and used large-scale survey methods. Our research, however, is qualitative in nature and we aimed for a small sample of teachers. Thus, our ideal sample was defined in our proposal as:

> We propose to work in two, possibly three, local authority areas with a sample of 24 teachers to be selected as follows:
>
> • All from primary schools (this is in the interests of continuity, since primary schools have been the focus of our researches for the past 12 years, and this study would be a natural extension of that work, as explained earlier).

- All to have been in their first post before 1988 and not older than 55 years of age.
- All to have experienced recent long-term absence from school for stress-related illness (this is our operational definition of stress for the purposes of this research).
- 12 men, 12 women.
- 12 headteachers, 12 classroom teachers
- Eight still receiving counselling, eight successfully returned to teaching, eight successfully adapted elsewhere.

This sample provides a clear focus (on primary and mid-career teachers), and promises rich chances for comparison among factors involving gender, position, and different forms of adaptation.

We propose to work collaboratively with a local authority Occupational Health Unit[2] which is currently engaged in counselling employees of the local authority (largely teachers, social workers and fire-service personnel) who are experiencing stress. The unit also has knowledge of those teachers who have returned to school or who have retired early or otherwise left teaching for stress-related reasons.

Teachers conforming to this definition were identified by the Occupational Health Unit which circulated our letter to these teachers, inviting them to take part in the research. The letter, in explaining the aims of the research, implied that we wished to do research *with* rather than *on* teachers and stated that the research findings would be used to give voice to teachers experiencing stress and also used to inform national and local policy with regard to stress in teaching. One intention of doing the research, therefore, is to change social situations for the better (Wolcott, 1995).

At this stage in the research 18 teachers have responded and of these 14 are women. Our sample, therefore, is predominantly composed of women. Some factors accounting for this composition might be as follows:

(a) The sample accurately reflects the proportion of women in the occupational group teachers.
(b) Women are more likely to suffer stress in their work.
(c) Women are more willing to report the experience of stress and, therefore, place themselves in situations demanding self-disclosure.
(d) Women are more able to self-disclose than men.
(e) Women are less prone to stigma as a result of stress than men.
(f) Willingness to be interviewed is a gendered manifestation of the confessional society and the 'modern confessional impulse' (Lee, 1993; Warren, 1988).
(g) Participants seek to publicize their plight and improve the situation for others.
(h) Respondents see the research interview as therapy.

The teachers who are taking part are being interviewed in their homes. Sikes and Measor (Sikes et al., 1985) also held their life history interviews in respondents' homes. The fact that many of the teachers they interviewed were men alone at home led to concerns about the safety of the female researchers.

Each interview is normally of one and a half to two hours duration. The intention is to have a number of interviews with each respondent over a two-year period. This adds a longitudinal dimension to the research.

In the following sections I analyze my gendered experience of in-depth longitudinal interviewing and field relations in terms of the researcher's self, rapport with respondents, power relations and the emotionality of the interviewer/interviewee relationship.

THE RESEARCHER'S SELF

> The crux of the issue is the interpretive moment as it occurs throughout the research process. The researcher brings considerable conscious and unconscious baggage into this moment, including: other related research; training within a particular discipline (such as anthropology); epistemological inclinations; institutional and funding imperatives; conceptual schemes about storytelling or power; social positionality (the intersection of race, class, gender, sexual orientation, among other key social locations); macrocultural or civilizational frames (including the research frame itself); and individual idiosyncracies, the interactions of which are themselves complex and ambiguous. This plethora of baggage in the guise of the interviewer, interacts with an interviewee, who, of course, brings her or his own baggage to the interaction. That the written result, the final interpretation of the interview interaction, is overloaded with the researcher's interpretive baggage is inevitable (Scheurich, 1995, p. 249).

I am white, male, middle aged, and consider myself to be working class, although most others who know me would say I was middle class (see Reay, 1996a, b, on this point). I was a school teacher in secondary and middle schools for over 20 years. During this time I conducted interactionist ethnographic research in the schools in which I worked. I have also taught in the Polytechnic/ University sector in teacher education and am currently an Associate Lecturer with the Open University. In these roles I worked with and came to know well a wide range of teachers in all sectors of education. My politics are of the left and I am critical of many of the educational 'reforms' which have taken place over the past two decades.

In my time in schools I had seen colleagues leave teaching owing to stress induced by the amalgamation of schools and national policy initiatives (for example, the raising of the school leaving age). More recently, I had completed a Ph.D study on the restructuring of primary teachers' work. This experience was not without its stresses and strains and during the research period I developed a stress-related illness. The research team I was working with at this time were producing data showing extensive teacher stress nationally in primary teaching (Woods, 1995); during Ofsted inspections (Jeffrey & Woods, 1998); and in the multi-cultural primary school (Boyle, 1994).

I had experience of stress during a year's unemployment when I had tried unsuccessfully to make a living as a supply teacher. My partner, an infant teacher, had suffered from stress-related illness in the past, is currently off work with stress and does not want to return to school. All of these biographical factors not only shaped the nature of the research topic and questions but also influenced the choice of methods and each stage of the research. For example, it was my partner's knowledge of and relationship with key personnel in the Occupational Health Unit which led me to a snowball sample of key informants. My empathy with respondents (see below) sprang directly from my own and my partner's experience of stress. My domestic situation informed the theories on stress I was generating and it fed directly into the questions I asked the respondents (see below).

In terms of 'body chemistry' I felt I would not constitute a 'sexual threat' to men or women respondents, as I felt rather like Douglas in that:

> I happen to be fortunate in being largely non-descript chemically . . . most people don't seem to notice me in a crowd – just one more 'normie', a middle aged man of medium height, slightly paunchy (these days), a bit bulbous headed and beady eyed (but not some extraterrestrial type) and a bit tanned and hacked in the face by time's bending sickle . . . I am not beautiful, sexy, otherwise exciting, or anything very distinct physically . . . I try to be fatherly-friendly, low profile but warm, very uncritical and very sympathetic and appreciative (Douglas, 1985, pp. 97–98).

The researcher, then, brings a great deal to the research enterprise and the act of interpretation:

> [H]ow researchers do this depends on the kind of self they bring to the interpretation – experiences undergone, interests and values, personal reference groups, affective disposition towards those studied, commitment to causes involved in the research (Woods, 1996, p. 54).

DEVELOPING TRUST AND RAPPORT

> Mutual trust is important and it is built up through dialogue not interrogation (Oakley, 1981, p. 44).

> Clergy wives became warm and eager to talk to me after the simple discovery that I was one of them (Finch, 1984, p. 79).

Normally, a vital first stage in the data collection process involves the researcher developing rapport with respondents. It is often argued that the greater the rapport the higher the quality of data which will be collected

(Wolcott, 1995; Woods, 1996). Interpersonal skill is required by the interviewer at this stage in order to develop a relationship with the interviewee in which even intimate secrets can be disclosed.

In the stress research this process began with an initial telephone conversation to set up an interview. The male respondents would negotiate details of the time and place of the interview and give route directions. In addition to this type of information some of the female respondents would enquire about the research and begin to start telling their stories. With many of the female respondents, then, rapport was beginning to be developed over the phone. I found, however, that relationships could also be broken in telephone conversations. When I phoned one woman to arrange a second interview her partner answered the phone and said that she was out but he would pass a message on. I did not hear from this respondent again. Either the woman was unwilling to take part in further interviewing or her partner had witheld his consent[3] for her to take part in the research (this latter possibility is raised by Chandler (1990)). A subsequent letter aimed at enlisting further support in the research was not answered by the respondent.

In the case of another respondent, I phoned on a Sunday afternoon and she was clearly preparing her lessons for school on Monday and seemed very tense and stressed. She said that she could not say that she could do another interview before discussing it with her husband. I recorded in my field notes that I was working on a Sunday and enjoying it yet she was doing the same but clearly not enjoying it. In addition to her schoolwork she also had domestic chores and the responsibilities of bringing up two young children. Consequently, she was juggling with the three shifts of work/children/home (Acker, 1994). In order for her to facilitate our first interview, her husband had to take the two children for a walk to create a 'space' for her to take part in the interview. They appeared at the end of it all soaking wet after trudging round the park in the rain for two hours. They feared coming back too early and disrupting the interview. She did not phone back and I guessed that as a family they were unwilling to participate further in the research, and again, I do wonder if the husband had withheld his permission for her to continue (Chandler, 1990).

While rapport is needed in order to collect qualitative data the researcher is cautioned against identifying too closely with respondents (Hammersley & Atkinson, 1995). What is recommended therefore is that the researcher knife-edges between *involvement* with respondents and maintaining analytical *distance* (Woods, 1996). The latter is needed for objectivity in research. In the first round of interviews I adopted an objectivist (Oakley (1981) would say objectivist masculinist) stance by starting each interview by giving brief details of the research aims, obtaining informed consent (when possible – see below) and noting some of the respondent's biographical details. This approach did not develop the kind of rapport I was seeking so I aimed to establish mutuality with respondents by informing them of my special characteristics (teaching, stress, maleness when interviewing men) and interest in the topic (personal and partner's stress). This involved me in self-disclosure and sharing my (and my partner's) experiences of stress. I obviously could not share the experience of being a woman with female respondents (Oakley, 1981), though I could that of a man with the men, by sharing, for example, my experience of conventionally male pursuits such as scouting with one respondent and rugby with another.

Initial meetings and interviews were the main occasions in which rapport was developed. There are advantages, of course in not developing rapport. Schutz' (1967) 'stranger' would commonly elicit more frank data and surprising openness than a 'friend'. This is usually the type of relationship existing between counsellor and counselled. I was told by respondents that their counsellors did not engage in self-disclosure. Some male and female respondents would say that they could not tell their story of stressful circumstances at work and experience of stress to colleagues, friends or family. Being a male researcher researching women could mean that the women did not expect me to share experiences as a woman and, therefore, would be more willing to elaborate on these aspects. A similar process occurred in multi-cultural research (Boyle, 1996) when young Islamic children explained in detail such events as worshipping in the Mosque to the Christian researcher. Their accounts, arguably, would have been less detailed and relied more on 'indexicality' (Goffman, 1981) with an Islamic researcher. Alternatively, women might not tell things that they would assume a man would not understand.

Offloading to a 'stranger' is seen by some (Brannen, 1988) to be a particular advantage of the 'one-off' interview. Brannen found that 'one-off' interviews on sensitive topics produced very high quality data because of the advantages of talking to a stranger and interviewees being comforted by the fact that they would never again 'cross paths' with the interviewer. By contrast, Oakley (1981) argues against the single interview because it 'encourages an ethic of detachment', and 'sisterly rapport' (developed over a long period) charac-terized her interviews. Laslett & Rappaport (1975) advocate longitudinal interviewing since better rapport increases the quality of the data. However, they stress the importance of the 'inter-interview dynamic'. In the first interview, they argue, the researcher is given a surface public account but in subsequent interviews accounts are 'more complete and deeper' (Lee, 1993). In the period between the first and second interview anxieties which are raised in the first interview generate resentments directed towards the interviewer. As long as interviewee and interviewer can handle this then it leads to richer data (Lee, 1993). I have a particularly vivid example of this from my research. A female respondent, who found recounting painful episodes in her first interview difficult, told me of this in her second interview. Her attitude towards me changed negatively in the course of this. It became apparent that throughout her life she had experienced difficulties with situations that demanded self-disclosure. She withdrew from the research following the interview. Fortunately for the research this has not been a reaction of the majority of participants. However, while women were easier to get to know and rapport developed with men and women over time, two other female respondents appeared to have developed Laslett & Rappaport's (1975) anxiety and dropped out of the research.

Developing rapport was also a feature of respondents' behaviour. They would welcome the researcher into their homes and prepare coffee, tea, or sometimes beer. Often drinks would be accompanied by biscuits or cakes. On some occasions after a long drive to the interview a meal would be offered. These type of acts signified the depth of 'sisterly' relationship in Oakley's (1981) research. I viewed them more as conventional 'manners' or acts of thoughtfulness rather than some kind of demonstration of gendered solidarity. Finch (1984) knew she had established rapport when respondents talked freely and at great length in answering her questions. This surprised her and was a function, she argued, of the woman-to-woman interview, in which mutuality was achieved as woman and clergyman's wife. I never encountered any difficulty in getting respondents to talk. I do not claim any superior abilities when it comes to interviewing but both the women and men in the study were very articulate and easy to talk to. I do not think there was a gender basis to this. All involved were teachers and, after all, earn their livings by communicating, usually by talking.

Interviews were friendly (though see later about challenging aspects) but respondents did not become my friends (Chandler, 1990) in the way that Cannon's (1989) respondents did. It was a kind of 'instrumental' relationship and, so far, I have developed no close attachments to the men and women in the research. The conversations I took part in, after all, did have a purpose (Burgess, 1988) other than developing rapport and friendship; I was there to collect data. Whether or not this could be considered as manipulative or exploitative is discussed below.

As rapport developed, the women (but not the men) would ask me how the research was going, how my partner was, and when she would be going back to work. The women in the research seemed to want to form a relationship which went beyond the interviewer. While this could, of course, be merely a further example of good manners, it seemed to me that it revealed a caring attitude absent in male respondents. The women said they had taken part in the research because they wanted to prevent what was happening to them happening to other teachers in the future. In turn, I felt obliged to give something back to respondents:

- Some women saw the interview as a therapeutic event and claimed it was just like counselling. Although I was not a trained counsellor, I did not consider I was causing harm in conducting the interviews. They did seem to be beneficial to most. One man has compared the interview situation to counselling and has benefited.
- I put one isolated (geographically and psychologically) woman in touch with a support group on workplace bullying.
- One woman who requested a transcript of our interview later used the text to help her confront painful experiences from the past and also act as a resource in cognitive therapy sessions.
- Some women wanted me to publish 'findings' to alert policy makers about the 'reality' of what was happening to teachers and schools.
- Some women wanted to see their stories documented and, through reading, relate their experiences to others in similar circumstances. It is this kind of rapport and respondent 'pressure' which led Patti Lather and Chris Smithies (1997) to desk-top publish their work (*Troubling the Angels: Women Living with HIV/AIDS*) and circulate it to respondents rather than wait for the publisher's slow process to be completed.

POWER RELATIONS

The event was still an interview; the purposes of the situation were not transformed by gender. The discussion was structured, purposeful conversation and I was structuring and, therefore, controlling it. The women's questioning of me was minimal compared to my questioning of them and their questions were different from the questions I asked. They looked for perfunctory biographical detail or sought reassurance on the normality of their feelings and experiences compared to other women who had been interviewed; I did not seek reassurance from them. Although the women asked me questions, they were not as interested in me as I was in them; the interviews were conducted in their homes not mine; they offered me hospitality, I did not offer it to them; I recorded the words of the interview while none of the women that I interviewed were, to my knowledge, recording me. The women I was interviewing were opening personal life to scrutiny and I was not (Chandler, 1990, p. 129).

The power dynamics of the research interview affect the quality of data collected. The power relationship between interviewer and interviewee has been described as equal when women interview women (Finch, 1984; Oakley,

1981). But normally the researcher/researched relationship is considered to be unequal, asymmetrical or in a state of imbalance. Oakley (1981) and Finch (1984) argued that women researchers and female interviewees share the same subordinated structural position in a male-dominated society. This egalitarian relationship (which logically men are unable to achieve with women) leads to greater rapport and better data (Lee, 1993). Critics of Oakley (see Wise, 1987), however, argue that she uses the "shared structural position of women as a magical device for the instant dissolution of inequalities" (Lee, 1993, p. 109). Even where gender differences are absent, imbalances of power can still exist between researcher and researched (Lee, 1993). This is particularly clear in the work of Reay (1996a, b); here it is social class not gender that is the perceived source of inequality. Although Reay just interviewed women she often occupied the subordinate social position in the research interview.

It is often assumed that the researcher holds the power in interviews particularly when men interview women. Given the substantive orientation of the research both male and female respondents could obviously be considered as vulnerable, since they were ill. The interview was far from an interrogation. I sometimes controlled the interview and followed the (memorized) semi-structured interview schedule. This was usual with male respondents. Some women would follow this routine but others would ask me, at the outset, how I wanted to go about the interview (for example a chronological narrative or critical incident approach). In these circumstances I gave them the choice of how to proceed. Some women, usually at the first interview, but some subsequently, would launch into their stories and seem most anxious to communicate everything quickly and in an apparently unstructured way. This not only took control away from me, but it also did not give me the opportunity of gaining informed consent for the interview and of explaining about confidentiality and anonymity. Oakley (1981) argues that in the survey interview conducted by men it is the (female) interviewee who is constructed as passive and the one who discloses. The male interviewer asks the questions and the female interviewee merely answers them. In my interviews, most often with women but sometimes with men, I would be asked questions. It was usually the women who would want to know how the research was going and how my partner was coping. They would also enquire about the plight of others in their position who were taking part in the research project. In these situations it was the women who had the power to make the male researcher self-disclose. And I always did, even though I might feel less obliged than they and had more power to refuse. Not to do this would, I felt, have not developed the relationship which in turn would impact negatively on the willingness of respondents to self-disclose. Additionally, I too, in certain circumstances, was using the interview as therapy.

Context is important in power dynamics, as anyone who has been told off in the headteacher's office will attest. The large majority of the stress interviews are being carried out in respondents' homes. This can, depending on the home, place them at an advantage in power-relations. Usually the men and women I interviewed shared my social class, although this was not always the case. Some respondents (male and female) were of a higher social class and enjoyed more material wealth than me. In these circumstances there was a power imbalance but it was not great enough to create the kinds of problems encountered in researching elites (Ball, 1994; Ozga & Gewirtz, 1994) but significant enough to create the kind of methodological difficulties Reay (1996a, 1996b) describes.

The researcher has power over respondents in that it is in her/his power to make the private account public. It was particularly important, therefore, to promise confidentiality and anonymity because stress is an illness potentially attracting stigma (Goffman, 1981). Also, a lot of the respondents' accounts I was listening to constituted 'whistleblowing' on bullying headteachers and managers. Informed consent was obtained where possible (see above). On only one occasion did I get asked (by a woman) for a signed statement guaranteeing confidentiality and anonymity. She was particularly concerned that she and her school should not be named for she was highly critical of the bullying treatment she had received from the headteacher. As a result of her experience at this school the respondent had lost trust in management and this, perhaps, had implications for her trust in researchers.

It is sometimes argued that qualitative research is manipulative or exploitative, that the researcher only enters the interviewer/interviewee relationship in order to get data and subsequently publish research reports. In this situation, the suffering of respondents is furthering the career of the researcher (Beynon, 1983). The 'humanising' interviews of feminist research, it is claimed, are non-exploitative (Chandler, 1990; Paterson, 1997). This position assumes that women researchers are not 'ambitious or competitive careerists' (Chandler, 1990) and that male researchers are. I know of no evidence to support this. In answer to a charge that I may have exploited the men and women participating in the research I can only point to the eventual use of research findings to improve teachers' working conditions and the personal help extended to some respondents.

The respondent has the power to define reality during the interview and interviewers disrupt this at their peril (Lee, 1993)! It is important to laugh at respondents' jokes and admire their pictures and furnishings, for example. This extends to suppressing one's own views even when in fundamental disagree-ment with the perspective of the respondent. For instance, a male respondent in the stress research made some remarks about Palestinians and the situation in Israel which I disagreed with and, in fact, found extremely offensive. However, I remained silent on this point, not wanting him to withdraw from the interview or the research project. Thus researcher silence became tacit agreement with the respondent in the same way that not intervening to protest at or stop an immoral act is to legitimate it.

In terms of power relations, respondents hold a great deal of power in that they can grant or withhold their consent to participate in the research. In longitudinal research on a 'sensitive topic' it is a constant fear for the researcher that a respondent will choose to withdraw from the project. Respondents, therefore, hold the ultimate sanction in their hands. No researcher welcomes high sample attrition rates.

EMOTIONS IN INTERVIEWS

> Researchers' fieldwork accounts typically deal with such matters as how the hurdles blocking entry were successfully overcome and the emergent relationships cultivated and maintained during the course of the research; the emotional pains of this work are rarely mentioned (Shaffir et al., 1980, pp. 3–4).

> Chances are that approaches and questions that make the researcher uncomfortable will have a similar effect on respondents (Wolcott, 1995, p. 108).

> Interviewing about sensitive topics can produce substantial levels of distress in the respondent which have to be managed during the course of the interview (Lee, 1993, p.105).

The research so far has established, perhaps unsurprisingly, that the experience of occupational stress involves the sufferer experiencing many negative emotions. Interviewing on the topic of stress, like other 'sensitive topics' (Lee, 1993) is potentially highly stressful itself for interviewee and interviewer alike. While the art of good listening is essential to the ethnographer (Wolcott, 1995), it is particularly important whilst interviewing in the area of occupational stress; an emotional investment in the research is demanded from the researcher. Respondents (both men and women) clearly wanted to offload and get worries off their chests. They were using the research interview for therapy. Interviews often took the form of counselling sessions with respondents not only engaging in self-disclosure but also rehearsing possible contributing factors for their illness. This usually involved respondents revealing their emotional responses to situations and their illness. Both men and women engaged in this form of emotional self-disclosure. Some men were reticent about self-disclosure just as some women were. Alternatively, some men and women would talk freely about their emotions. Indeed many had been encouraged to do so previously in counselling sessions.

The research interview as counselling involves the processes of 'transference' and 'counter-transference'. These concepts are derived from psychoanalytic theory and 'transference' refers to "feelings derived from earlier experiences which are projected onto the analyst." " 'Counter-transference' refers to similar feelings on the part of the clinician" (Lee, 1993, p. 105). Thus, my emotional experiences of stress were inevitably relived in interview. I, therefore, had to cope with not only the respondent offloading but myself having to revisit painful experiences.

Of significance here was my response to the expression of deep emotions by respondents. The emotional reactions in interview varied between men and women. During interviews with some of the women their eyes would fill with tears or they started crying at some stage; this never happened in interviews with the men (though this was a feature of a female colleague's [Carlyle, this volume] interviews with some men). While crying is only one way of manifesting distress it is highly visible and can elicit a response in others. One interviewee, for example, 'broke down' and had to leave the room in which the interview was taking place (her sitting room) and go to another room to recover. For a few minutes I could hear the sounds of nose-blowing and sobbing from the next room. During such episodes I experienced acute embarrassment and I found that the only way I could help was to proffer some tissues when the respondent returned. I no doubt also gave non-verbal expressions of sympathy and changed my tone of voice. In the remainder of the interview I was anxious not to return to the issue which had caused this response. In avoiding this and closely associated areas I was, of course, failing to gather much potentially useful data on the emotions and occupational stress (see below).

At the end of interviews I would thank respondents for their willingness to put themselves in a situation where revisiting painful episodes was likely.

In terms of interviewers handling distress in the research interview Lee summarizes Brannen's (1988) argument that:

> [F]aced with such distress interviewers may want to help but should strongly question their motives for doing so. Such feelings on the part of the interviewer, she suggests, "often have more to do with helping the helper than those who are in need" (1988, p. 559). All that may be possible in these situations is for the interviewer to undertake the difficult task of enduring and sharing the pain of the respondent (Lee, 1993, p.106).

In my particular incident described above the question which had precipitated this response was about the woman's sense of self-esteem during her illness. Her refusal/inability to articulate a verbal response to this question might be viewed as a 'resistance' in a psychodynamic counselling interview (Jacobs, 1988), when the respondent refuses to disclose their feelings. Following such 'resistances' I would move on to a 'safer' topic and avoid mention of the area which had caused difficulty. It seems to me that to pursue a line of questioning which has caused obvious pain and distress would be akin to asking the victim of a road crash (at the scene of the accident) how they felt about the crash. I do not know whether to attribute this reaction on my part to my maleness, my lack of training as a counsellor or as a means of coping with my stress in the interview. Like Chandler (1990) I felt that:

> Counsellors would not set out to rake through people's biographies and pry into what may
> be tragic memories and human despair without some training and clear guidelines for their
> behaviour, but researchers do (Chandler, 1990, p.131).

Counsellors have strategies for enabling people to self-disclose and calculate whether it is beneficial or harmful for the client to do so. Of course, the longitudinal interview method used in this research might prove to be one strategy for encouraging progressive self-disclosure. Other strategies are derived from counselling. For example, a female colleague (see Carlyle, 1998 and this volume) whom I have observed interviewing on the topic of stress is far more successful in exploring male and female respondents' emotional states. However, she attributes this difference not only to her gender and her life experience of emotional labour but also to her background in person-centred humanistic counselling.

Interviewing on stress exacts its toll on interviewer (male and female) and interviewee (male and female) alike. But at the end of interviews I would usually feel drained and the interviewee would want to continue. The colleague I mentioned earlier seeks counselling (with a female counsellor) herself as one strategy in dealing with emotions and offloading some of the feelings of guilt and anger projected by respondents and accumulated in the interview situation. For myself, I have not received counselling. Apart from my avoidance of intense emotional states in interviews which I have described, I consider that I achieve emotional and analytical distance by writing reflexively (eg. field notes and analytical memos) about my experiences researching stress. Perhaps adopting this independent, non-involved, 'objectivist' stance might be a particularly 'masculine' thing to do, but it is nonetheless necessary to achieve the scientific view.

CONCLUSION

> Feminism's keen sensitivity to structural inequalities in research and to the irreconcilia-
> bility of Otherness applies primarily, I believe, to its critique of research by men,

particularly to research by men, but about women. The majority of feminist claims about *feminist* ethnographic and other forms of qualitative research, however, presume that such research occurs almost exclusively woman-to-woman. Thus, feminist researchers are apt to suffer the delusion of alliance more than the delusion of separateness . . . (Stacey, 1991, p. 116).

In the years following Oakley's (1981) critique, researchers (both female and male) have become more sensitive to issues of gender in qualitative methodology. This trend, in part, explains the focus and style of this chapter. In my reflexive account I have tried to examine critically my fieldwork experiences in research on a sensitive topic in order to identify some of the strengths and weaknesses of a male researcher undertaking qualitative research involving female and male respondents. Some of the advantages and disadvantages of this approach have been discussed. All of these had implications for the type and quality of data which have been elicited in the stress research. However, I have not been alone in my research endeavours. My membership of a research team has been invaluable in developing constant comparative work (Glaser & Strauss, 1967). In team meetings and informal conversations I had the opportunity of engaging in methodological discussions and shared data analysis with the female colleague mentioned earlier who is engaged in qualitative work on stress with secondary teachers. Here, her perspectives and experiences in the field with male and female respondents have informed my subsequent fieldwork strategies, relationships and analyses. Of course, in our joint analyses of transcripts, fieldnotes and memos, we are not deriving the same insights from the same data. As stated at the outset of this chapter, qualitative research and interpretation is a highly individual process for the researcher. The researcher's self is invested in each stage of the research activity. Gender is only one facet of that self.

NOTES

1. Woods (1990) has criticized the 'objective', 'cold-eyed', cold science (Apple, 1986) masculinist studies of women teachers (see Waller, 1932 and Lortie, 1975). He advocates that in place of this approach such studies need to embrace the 'warm hearted' subjectivity of feminine scholarship.

2. It was the female manager of the Occupational Health Unit who acting as a 'gatekeeper' asked me (while negotiating access) if I could handle interviews sympathetically and inquired if I had had any training in counselling techniques. Many of the respondents, she warned, were "very angry and very upset at what had happened to them."

3. Quite apart from cultural norms and power relations which might be in play here the situation I describe could also have been influenced by stress-induced illness. One finding of the stress research is that the initial stage in the illness of some participants was characterized by physical and emotional exhaustion. In this state sufferers would

find it extremely difficult to make decisions, even of the most mundane kind. Those participants who had partners would, therefore, delegate responsibilities in terms of decision-making until they had recovered. Alternatively, some men/women would recognize this aspect of the illness and assume responsibilities in order to be protective and 'shield' their partners from external intrusion. Both males and females behaved in this way.

REFERENCES

Acker, S. (1994). *Gendered Education: Sociological Reflections on Women, Teaching and Feminism*. Milton Keynes: Open University Press.
Aoki, T. T. (1983). Experiencing Ethnicity as a Japanese Canadian Teacher: Reflections on a Personal Curriculum. *Curriculum Inquiry, 13*(3), 321–335.
Apple, M. (1986). *Teachers and Texts: a Political Economy of Class and Gender Relations in Education*. New York: Routledge and Kegan Paul.
Ball, S. J. (1994). Researching Inside the State: Issues in the Interpretation of Elite Interviews. In: D. Halpin & B. Troyna (Eds), *Researching Education Policy: Ethical and Methodological Issues*. London: Falmer.
Beynon, J. (1983). Ways-in and Staying-in: Fieldwork as Problem Solving. In: M. Hammersley (Ed.), *The Ethnography of Schooling*. Driffield: Nafferton.
Boyle, M. (1994). Child-Meaningful Learning in a Bilingual School. ESRC Award R000235123, (1994–7).
Boyle, M. (1996). *Exploring the Worlds of Childhood: The Dilemmas and Problems of the Adult Researcher*, Paper presented at the Ethnography and Education Conference, Oxford, September.
Boyle, M., & Woods, P. (1996). The Composite Head: Coping with Changes in a Primary Headteacher's Role. *British Educational Research Journal, 22*(5), 549–568.
Brannen, J., & Collard, C. (1982). *Marriages in Trouble: The Process of Seeking Help*. London: Tavistock.
Brannen, J. (1988). The Study of Sensitive Subjects. *Sociological Review, 36*, 552–563.
Burgess, R. G. (Ed.) (1984). *The Research Process in Educational Settings: Ten Case Studies*. Lewes: Falmer Press.
Burgess, R. G. (1988). Conversations with a Purpose: the Ethnographic Interview in Educational Research. In: R. G. Burgess (Ed.) *Studies in Qualitative Methodology 1*. London: JAI.
Cannon, S. (1989). Social Research in Stressful Settings: Difficulties for the Sociologist Studying the Treatment of Breast Cancer. *Sociology of Health and Illnes, 11*(1), 62–77.
Carlyle, D. (1998). Opening the Can of Worms: gender and emotion in sensitive research. Paper presented at the Ethnography and Education Conference, Oxford, September.
Chandler, J. (1990). Researching and the Relevance of Gender. In: R. G. Burgess (Ed.) *Studies in Qualitative Methodology 2*. London: JAI.
Currer, C. (1992). Strangers or Sisters? An Exploration of Familiarity, Strangeness, and Power in Research. In: R. G. Burgess (Ed.) *Studies in Qualitative Methodology 3*. London: JAI.
Davies, B. (1992). Women's Subjectivity and Feminist Stories. In: C. Ellis & M. G. Flaherty (Eds), *Investigating Subjectivity: Research on Lived Experience*. London: Sage.
Delamont, S. (1997). Fuzzy Borders and the Fifth Moment: Methodological Issues Facing the Sociology of Education. *Review Essay, British Journal of Sociology of Education, 18*(4), 601–606.

Denzin, N. K. (1997). *Interpretative Ethnography: Ethnographic Practices for the 21st Century.* Thousand Oaks, CA: Sage.

Douglas, J. (1985). *Creative Interviewing.* Newbury Park, CA: Sage.

Ellis, C., & Flaherty, M. G. (Eds) (1992). *Investigating Subjectivity: Research on Lived Experience.* Newbury Park: Sage.

Finch, J. (1984). It's Great to Have Someone to Talk to: the Ethics and Politics of Interviewing Women. In: C. Bell & H. Roberts (Eds), *Social Researching: Politics, Problems, Practice.* London: Routledge and Kegan Paul.

Fleuhr-Lobban, C., & Lobban, R. A. (1986). Families Gender and Methodology in the Sudan. In: T. L. Whitehead & M. E. Conaway (Eds), *Self, Sex and Gender in Cross-Cultural Fieldwork.* Urbana: University of Illinois Press.

Freedman, D. (1986). Wife, Widow, Woman: Roles of an Anthropologist in a Transylvanian Village. In: P. Golde (Ed.), *Women in the Field: Anthropological Experiences*(2nd ed.). Berkeley: University of California Press.

Friedl, E. (1986). Fieldwork in a Greek Village. In: P. Golde (Ed.), *Women in the Field: Anthropological Experiences* (2nd ed.). Berkeley: University of California Press.

Glaser, B. G., & Strauss, A. L. (1967). *The Discovery of Grounded Theory.* Chicago: Aldine.

Goffman, E. (1963). *Stigma: Notes on the Management of Spoiled Identity.* London: Penguin.

Goffman, E. (1981). *Forms of Talk.* Oxford: Basil Blackwell.

Hammersley, M., & Atkinson, P. (1995). *Ethnography: Principles in Practice* (2nd ed.). London: Routledge.

Hey, V. (1997). *The Company She Keeps: an Ethnography of Girls' Friendship.* Buckingham: Open University Press.

Hughes, C. (1992). A Stranger in the House: Researching the Stepfamily. In: R. G. Burgess (Ed.), *Studies in Qualitative Methodology 2.* London: JAI.

Jacobs, M. (1988). *Psychodynamic Counselling in Action.* London: Sage.

Jeffrey, B. (1998). Problems of Gender Conceptualizations in Current Educational Research on Teachers. *Analytical Memo*, February. The Open University.

Jeffrey, B., & Woods, P. (1998). *Teachers under Inspection: the Impact of Inspections on Primary Teachers and their Work.* London: Falmer.

Johnson, W. T., & P. Delamater, (1976). Response Effects in Sex Surveys. *Public Opinion Quarterly, 40*, 165–181.

Laslett, B., & Rappoport. R. (1975). Collaborative Interviewing and Interactive Research. *Journal of Marriage and the Family, 37*, 968–977.

Lather, P. (1997). Drawing the Line at Angels: Working the Ruins of Feminist Ethnography. *Qualitative Studies in Education, 10*(3), 285–304.

Lather, P., & Smithies, C. (1997). *Troubling the Angels: Women Living with HIV/AIDS.* Boulder, CO: Westview.

Lee, R. M. (1993). *Doing Research on Sensitive Topics.* London: Sage.

Lortie, D. C. (1975). *Schoolteacher.* Chicago: University of Chicago Press.

Nader, L. (1986). From Anguish to Exultation. In: P. Golde (Ed.), *Women in the Field: Anthropological Experiences.* Berkeley: University of California Press.

Oakley, A. (1979). *Becoming a Mother.* Oxford: Martin Robertson.

Oakley, A. (1981). Interviewing Women: a Contradiction in Terms? In: H. Roberts (Ed.), *Doing Feminist Research.* London: Routledge and Kegan Paul.

Ozga, J., & Gewirtz, S. (1994). Sex, Lies and Audiotape: Interviewing the Education Policy Elite. In: D. Halpin & B. Troyna (Eds), *Researching Education Policy: Ethical and Methodological Issues.* London: Falmer.

Paterson, A. S. (1997). A Humanistic Framework for Interviewer Skills. Paper presented to the British Educational Research Association Conference, University of York, September.

Reay, D. (1996a). Taking the Words out of Women's Mouths: Interpretation in the Research Process. *Feminist Review, 53* (Summer), 57–73.

Reay, D. (1996b). Dealing with Difficult Differences: Reflexivity and Social Class in Feminist Research. *Feminism and Psychology, 6*(3), 443–456.

Scheurich, J. J. (1995). A Postmodernist Critique of Research Interviewing. *Qualitative Studies in Education, 8*(3), 239–252.

Schutz, A. (1967). *Collected Papers.* The Hague: Nijhoff.

Shaffir, W., Marshall, V., & Haas, J. (1980). Competing Commitments: Unanticipated Problems of Field Research. *Qualitative Sociology, 2*, 56–71.

Sikes, P. J., Measor, L., & Woods, P. (1985). *Teacher Careers: Crises and Continuitites.* Lewes: Falmer.

Stacey, J. (1991). Can There be a Feminist Ethnography? In: S. Gluck & D. Patai (Eds), *Women's Words: The Feminist Practice of Oral History.* London: Routledge.

Sudarkasa, N. (1986). In a World of Women: Fieldwork in a Yoruba Community. In: P. Golde (Ed.), *Women in the Field: Anthropological Experiences.* Berkeley: University of California Press.

Troman, G. (1997). The Effects of Restructuring on Primary Teachers' Work: A Sociological Analysis. Unpublished Ph.D. thesis, The Open University.

Walford, G. (Ed.) (1991). *Doing Educational Research.* London: Routledge.

Waller, W. (1932). *The Sociology of Teaching.* New York: Wiley.

Warren, C. A. B. (1988). *Gender Issues in Field Research.* Newbury Park, CA: Sage.

Wise, S. (1987). A Framework for Discussing Ethical Issues in Qualitative Research: a Review of the Literature. *Studies in Sexual Politics, 19*, 47–88.

Wolcott, H. F. (1995). *The Art of Fieldwork.* London: Altamira.

Woods, P. (1986). *Inside Schools: Ethnography in Schools.* London: Routledge.

Woods, P. (1990). Cold Eyes and Warm Hearts: Changing Perspectives on Teachers' Work and Careers. *British Journal of Sociology of Education, 11*(1), 101–117.

Woods, P. (1995). The Intensification of the Teacher's Self. Paper presented at the Conference on Teacher Burnout, Marbach, November.

Woods, P. (1996). *Researching the Art of Teaching: Ethnography for Educational Use.* London: Routledge.

Woods, P., Jeffrey, B., Troman, G., & Boyle, M. (1997). *Restructuring Schools, Reconstructing Teachers: Responding to Change in the Primary School.* Buckingham: Open University Press.

NOTES ON CONTRIBUTORS

Denise Carlyle is a PhD student in the Centre for Sociology and Social Research in the School of Education at The Open University. Her research focuses on teacher emotion and stress-related illness. She has worked in clinical audit, and as a sixth-form teacher of psychology. Key interests include gender, identity and mental health, 'care in the community', the sociology of emotion and the sociology of health and illness.

Tuula Gordon is Professor of Women's Studies and Social Sciences in the Department of Women's Studies, University of Tampere. She obtained her Sociology degree from the London School of Economics and her doctorate from the University of London. She has also taught in the University of Helsinki. Publications include: *Democracy in One School? Progressive Education and Restructuring* (1986, Falmer Press), *Feminist Mothers* (1990, Macmillan), *Single Women: On the margins?* (1994, Macmillan,) *Making Spaces: Citizenship and Difference in Schools* (with Janet Holland and Elina Lahelma, 2000, Macmillan).

Valerie Hey is a senior researcher in the academic group, Culture, Communication and Societies, Institute of Education, University of London. She has published widely in the fields of gender, education, cultural studies and feminist methodology. Her acclaimed study *The Company She Keeps: an ethnography of girls' friendship* (1997) brought a feminist poststructuralist understanding to the formal and informal domains of schooling. Her most recent writing takes up arguments about identity, identifications and sub-jectivity and brings these questions together in developing critiques of New Labour's construction of 'community'. She is particularly interested in theorising 'social capital' and 'network poverty' in the context of school and community relations. She is currently co-directing an Economic and Social Research Council study, Learning and Gender, which looks at the relations of pedagogy, gender and achievement in primary schools.

Janet Holland is Professor of Social Research and Director of the Social Sciences Research Centre at South Bank University, London. Publications include: *The Male in the Head* (with Caroline Ramazanoglu, Sue Sharpe and

231

Rachel Thomson, 1998, Tufnell Press); *Sex, Sensibility and the Gendered Body* (edited with Lisa Adkins) and *Sexual Cultures* (edited with Jeffrey Weeks, both 1996, Macmillan); *Debates and Issues in Feminist Research and Pedagogy* and *Identity and Diversity: Gender and the Experience of Education* (both edited with Maud Blair, 1995, Multilingual Matters with The Open University), *Making Spaces: Citizenship and Difference in Schools* (with Tuula Gordon and Elina Lahelma, 2000, Macmillan).

Caroline Hudson is a Research Officer at the Department of Educational Studies, University of Oxford. Her current research interests are two-fold. She is researching teachers' implementation of a range of strategies to improve students' writing in subjects across the curriculum. She is also evaluating a programme of preventative policing in five Oxfordshire secondary schools. Her doctorate explored young people's perceptions of the relationships between their family structure and their experiences of family and schooling. Her previous research has also included student attendance and truancy in an Oxfordshire secondary school; the views of students in five Oxford upper schools of changing from a three-tier to a two-tier model of schooling; and an evaluation of a literacy intervention for children in care.

Mary Jane Kehily has been actively involved in the teaching of Personal and Social Education and has research interests in sexuality, schooling and popular culture. After completing a graduate course at the Department of Cultural Studies, University of Birmingham, she worked as a researcher at the University of Birmingham and the Centre for Educational Development, Appraisal and Research (CEDAR), University of Warwick. She is currently working on an ESRC funded project, Children's Relationship Cultures in Years 5 and 6, based at the Institute of Education, University of London.

Elina Lahelma is Docent in Sociology of Education in the Department of Education, University of Helsinki. Currently she works as Senior Fellow at the Academy of Finland. She has written extensively on gender and education, including *Making Spaces: Citizenship and Difference in Schools* (with Tuula Gordon and Janet Holland, 2000, Macmillan).

Anoop Nayak is a lecturer in cultural geography at the University of Newcastle upon Tyne. Previously he studied at the Department of Cultural Studies at the University of Birmingham and has taught at Liverpool, John Moore's University. He has research interests in the following areas: race and ethnicity; ethnography; masculinities; and youth subculture. For the last year, Anoop has been working in the Department of Education, University of Newcastle upon Tyne, on a funded research project exploring the biographical experiences of ethnic minority student teachers and their transition into the workplace. His current research focus is on social history, whiteness and the meaning of white ethnicity within young people's lives.

Andrew Parker is a lecturer in sociology/CEDAR at the University of Warwick. He was previously a Senior Lecturer in the School of Behavioural Studies at Nene University College, Northampton. He has written about the construction of masculinity in various educational and sporting locales and is an ex-physical education teacher.

Jayne Raisborough is an ESRC funded research student at Lancaster University. She is currently completing a doctorate which explores the discursive borders around women's participation in a uniformed youth organisation. She is also an associate lecturer at Liverpool Hope University College where she lectures in feminism, medical sociology, leisure, and social research. Her further research interests change weekly but all reflect her fascination with the subtleties of gender and sexuality across a number of sites. She lives in the North-west with her partner, aggressive cat and doting dog.

Christine Skelton is a lecturer in the Department of Education at the University of Newcastle. She is Director of Undergraduate Programmes and teaches gender and education modules on undergraduate and postgraduate courses. Her PhD research was a case study of masculinities in primary schooling undertaken from a feminist perspective. She has published widely from this research which explored issues relating to men teachers and boys. She is currently writing a book *Schooling the Boys: Masculinities and Primary Education* (to be published by Open University Press).

Geoff Troman is a Research Fellow and Associate Lecturer in the School of Education at the Open University. Geoff taught science for twenty years in secondary modern, comprehensive and middle schools before moving into higher education in 1989. Throughout his time in schools he carried out research as a teacher researcher. His recently completed doctorate is an ethnography of primary school restructuring. He is currently conducting research on the social construction of teacher stress. Among other publications in the areas of qualitative methods, school ethnography and policy sociology, he co-authored *Restructuring Schools, Reconstructing Teachers* (with Peter Woods, Bob Jeffrey and Mari Boyle, Open University Press, 1997).

Geoffrey Walford is Professor of Education Policy and a Fellow of Green College at the University of Oxford. He was previously Senior Lecturer in Sociology and Education Policy at Aston Business School, Aston University, Birmingham. His recent books include: *Choice and Equity in Education*

234

(Cassell, 1994), *Educational Politics: Pressure groups and faith-based schools* (Avebury, 1995), *Affirming the Comprehensive Ideal* (Falmer, 1997, edited with Richard Pring), *Doing Research about Education* (Falmer, 1998, editor) and *Durkheim and Modern Education* (Routledge, 1998, edited with W. S. F. Pickering). His research foci are the relationships between central government policy and local processes of implementation, choice of schools, religiously-based schools and ethnographic research methodology. He is joint-editor of the *British Journal of Educational Studies* and is currently directing a Spencer Foundation funded comparative project on faith-based schools in England and The Netherlands.